Voltammetric Determination of Molecules of Biological Significance

Voltammetric Determination of Molecules of Biological Significance

W. FRANKLIN SMYTH
Department of Applied Physical Sciences
University of Ulster, Coleraine,
Northern Ireland

JOHN WILEY & SONS
Chichester · New York · Brisbane · Toronto · Singapore

Other Wiley Editorial Offices

John Wiley & Sons, Inc., 605 Third Avenue,
New York, NY 10158-0012,USA

Jacaranda Wiley Ltd, G.P.O. Box 859, Brisbane,
Queensland 4001, Australia

John Wiley & Sons (Canada) Ltd, 22 Worcester Road,
Rexdale, Ontario, M9W 1L1, Canada

John Wiley & Sons (SEA) Pte Ltd, 37 Jalan Pemimpin #05-04,
Block B, Union Industrial Building Singapore 2057

Library of Congress Cataloging-in-Publication Data

Smyth, W. Franklin.
Voltammetric determination of molecules of biological significance
W. F. Smyth.
p. cm.
Includes bibliographical references (p.) and index.
ISBN 0 471 93345 7
1. Biomolecules—Analysis. 2. Voltammetry. I. Title.
QP519.9.E42S68 1992
574.19′285—dc20 91-36594

British Library Cataloguing in Publication Data

A catalogue record for this book is available from the British Library

ISBN 0 471 93345 7

Typeset set in Times 10/12pt by APS, Salisbury, Wilts
Printed and bound in Great Britain by Biddles Ltd, Guildford, Surrey

Dedication

This book is dedicated to the memory of my late parents, William and Anna Smyth, to my wife Jane and daughters Annya, Amy, Clare and Frances.

Contents

Preface

Voltammetric techniques of analysis, i.e. those based on the measurement of current resulting from an oxidation or reduction at an electrode surface following the application of a potential difference to an electrochemical cell, have assumed an important place in the arsenal of analytical techniques for the identification and determination of trace concentrations (i.e. 10^{2}–10^{-4} ppm) of many organic, organometallic and inorganic molecules of biological significance. Techniques such as differential-pulse polarography (DPP), square-wave voltammetry (SWV), anodic, cathodic and adsorptive stripping voltammetry (ASV, CSV and AdSV), electrochemical detection coupled with flow-injection analysis (FIA–ED), ion chromatography (IC–ED) and high-performance liquid chromatography (HPLC–ED), electrochemical immunoassay and the application of amperometric biosensors are particularly important in this regard. Organic and organometallic molecules of biological importance as pollutants, drugs, food additives and contaminants, neurotransmitters, etc., as they are contained in a variety of complex matrices such as air, water, soil, plasma and tissue, can be determined by these voltammetric techniques at trace concentrations usually after the application of sample pretreatment procedures and certain separation techniques. In some cases, it is possible to determine the molecule directly in the matrix when the electrochemical reaction is particularly selective in the presence of potential interferents. Certain environmentally relevant inorganic molecules, such as the oxyanions of transition elements, can be determined by polarography, stripping voltammetry or on-line ED following either IC or HPLC. This text is a critical state-of-the-art discussion of recent applications of voltammetry to the identification and determination of selected molecules of biological significance. It is subdivided where possible according to the nature of the electroactive functional groups present in the overall molecular structures and follows the order nitrogen-, sulphur- and oxygen-containing molecules, hydrocarbons, halogen-containing and organophosphorus molecules, organometallics and selected inorganic molecules. Wherever possible, a critical comparison is made between these voltammetric methods and competing analytical methods based primarily on chromatographic and spectroscopic techniques. The reader is referred to standard texts for information on the basic principles of the particular voltammetric techniques, e.g. modern polarographic techniques (Bond, 1980), stripping voltammetry (Smyth, 1987), electrochemical detection (Palmisano and Zambonin, 1984), electrochemical immunoassay (Heineman and Halsall, 1985) and amperometric biosensors (Frew and Hill, 1987).

My thanks go to our departmental secretary, Carol Millar, for typing the text, and to my daughter Frances who helped prepare the index.

W. FRANKLIN SMYTH

1

Voltammetric Analysis of Selected Nitrogen-containing Molecules

1.1 MOLECULES CONTAINING THE NITRO GROUP

Aliphatic compounds such as nitramines, containing the N—NO_2 functional group, and nitrate esters, with the —O—NO_2 group, are commonly encountered as explosives. Detection of explosives has been the subject of investigations from a variety of directions. One aspect is environmental hygiene in the manufacturing plant, e.g. traces of ethylene glycol dinitrate (EGDN) have been found in the biological fluids of workers involved in its production (Williams and Murray, 1966). In addition, forensic scientists have examined post-blast debris in search of explosive material and also gunshot residues in order to determine whether a suspect had fired the weapon in question or not. Further, these explosive compounds can contaminate local natural waters either from the plant effluents or following the explosive blast.

Bratin *et al.* (1981) investigated the reduction behaviour of candidate explosive compounds and have proposed that nitrate ester groups reduce stepwise in $2e^-$ processes as follows:

$$RONO_2 + 2e^- + H_2O \rightarrow ROH + OH^- + NO_2^-$$

Nitramines such as RDX, which contains three N—NO_2 groups in a six-membered heterocyclic molecule, are proposed to be reduced in two steps as follows:

$$\text{RDX} + 6H^+ + 6e^- \rightarrow N\text{-nitroso intermediate} \xrightarrow[+12e^-]{+12H^+} \text{parent amine}$$

They used reversed-phase HPLC [Biophase C_{18} column, 0.02 M monochloroacetic acid–0.0147 M sodium acetate–0.001 M EDTA–5% (v/v) ethanol–17% (v/v) propan-1-ol pH 3.5 mobile phase at a flow rate of 1.7 ml min^{-1}] with reductive electrochemical detection at an Au/Hg electrode held at −1.00 V (vs Ag/AgCl) to separate and determine nitrate esters, nitramines and nitroaromatic explosive compounds as shown in Table 1. Detection limits in all cases for a signal-to-noise ratio (S/N) of 3 were lower than for UV detection at 254 nm. For the nitrate ester, nitroglycerin, the Au/Hg amperometric detector gave an almost 500 times lower detection limit than the UV detector.

An HPLC method was then developed for the detection of organic gunshot residues on the hands of individuals who had discharged a weapon. Identification was focused on nitroglycerin and 2,4-dinitrotoluene (2,4-DNT) by reductive ED and diphenylamine (DPA) by oxidative ED at a glassy carbon indicator electrode. The analytical method was relatively simple in that, after the revolver had been fired, the residue was collected from the back of the firing hand with a cotton swab dipped in acetone, and the swab was washed with 20% ethanol in water solution. This was then filtered through a 0.2 μm nitrocellulose filter and finally 20 μl of the deoxygenated solution were injected on to the HPLC column. For those smokeless gunpowders which contained nitroglycerin and 2,4-DNT, these were detected in the swabbed sample after firing, which was indicative of firing the appropriate weapon or handling nitroglycerin-based explosives. DPA occurred in the tested smokeless gunpowders at levels of the order of 0.5 g per 100 g and was also detected in residues using oxidative ED at a glassy carbon electrode. Such chromatograms were more complex than the reductive-mode chromatograms owing to the interference of phenolic compounds present in the lipid layer in the skin, especially if handcream lotions containing parabens were used.

Maskarinec *et al.* (1984) also used reductive ED following HPLC to determine these explosive compounds in natural waters. The analytical method involved adsorption of the molecules from aqueous samples on Porapak resin, followed by desorption with acetone and measurement of the nitro compounds

Table 1. Summary of liquid chromatographic results on a C_{18} non-polar stationary phase for selected explosive compounds

		Detection limit (ng) at S/N = 3	
Compound	*k'*	UV at 254 nm	ED at − 1.0 V
Nitroglycerin	3.96	160	0.38
2,4-DNT	4.96	0.57	0.16
2,6-DNT	5.00	1.2	0.17
3,4-DNT	5.50	1.3	0.15
PETN	7.90	—	0.40
HMX	0.21	1.5	0.29
Picric acid	0.58	0.47	0.065
RDX	0.88	0.88	0.17
Tetryl	2.58	0.77	0.21
TNT	3.04	0.65	0.14

by HPLC–ED with an Au/Hg electrode held at -1.00 V (vs Ag/AgCl), with detection limits in the low ppb region. The selectivity of this technique is illustrated in Fig. 1 for the separation of six explosive compounds at the ppm level, with the ED showing superior performance over the UV detector at 210 nm.

Krull *et al.* (1984) exploited the technique of HPLC–photolysis–oxidative ED for the determination of these organonitro compounds that cannot be oxidised as the parent molecules. NO_2^- can be released from these compounds by post-column on-line photolysis and this entity detected at potentials $\geqslant +0.8$ V (Fig. 2). They have found detection limits of 25 ppb for RDX, TNT and Tetryl and 200 ppb for nitroglycerin. They also investigated the effect of inorganic and

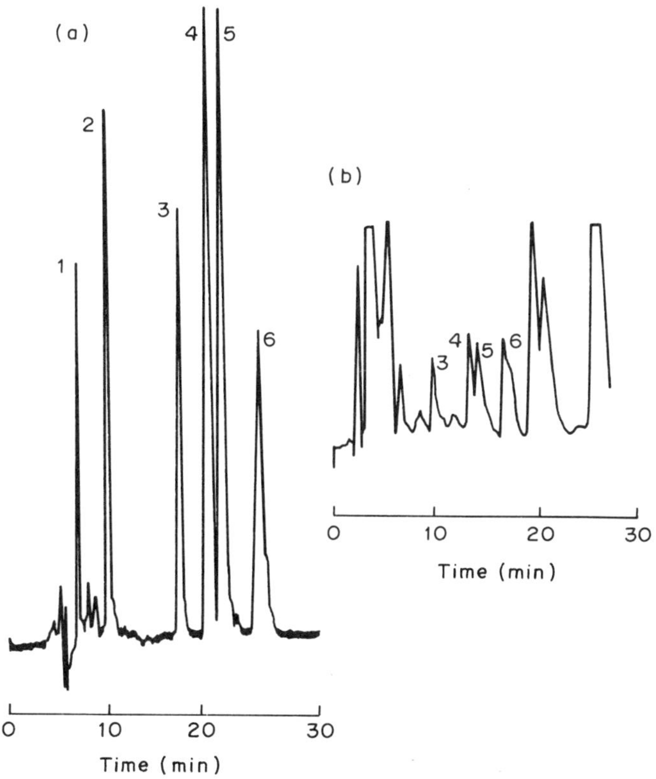

Figure 1. Detection of munitions following adsorption and desorption using Porapak R. (a) ED, 100 nA full-scale; injection volume, 20 μl. (b) UV, 210 nm, 0.02 a.u.f.s; injection volume, 50 μl. Peaks: 1 = HMX (8 μg ml^{-1}); 2 = RDX (2 μg ml^{-1}); 3 = NG (4 μg ml^{-1}); 4 = 2,6-DNT (2 μg ml^{-1}); 5 = 2,4-DNT (2 μg ml^{-1}) and 6 = PETN (6 μg ml^{-1}). Acknowledgement is made to the Oak Ridge National Laboratory operated by Martin Marietta Energy Systems for the US Department of Energy, for permission to reproduce this Figure. Also reproduced by permission of Elsevier Science Publishers

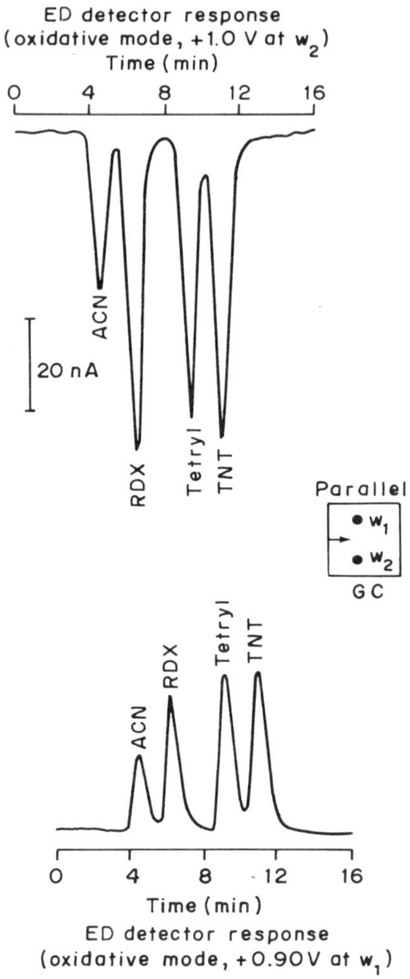

Figure 2. HPLC-*hν*-ED dual electrode chromatograms of RDX, Tetryl and TNT using a C-18 reversed-phase 3 μm, 10 cm × 4.6 mm i.d. column with methanol–0.1 M NaCl (1 + 1) at 0.6 ml min^{-1} flow rate. Dual electrodes operated in the oxidative/oxidative modes using glassy carbon surfaces. From Krull *et al.* (1984). Copyright ASTM. Reprinted with permission

organic anionic interferences and found that the anions I^-, IO_3^-, IO_4^-, HCO_3^-, NO_2^-, NO_3^-, HSO_3^-, S^{2-}, benzoate, CNS^-, $H_2PO_4^-$ and HPO_4^- were electroactive at + 0.8 and + 1.0 V using the technique of FIA-*hν*-ED with a dual glassy carbon electrode detection system, operating in the parallel mode.

Aromatic nitro compounds are encountered widely as molecules of biological significance and can be determined by a variety of voltammetric techniques (Table 2). Where the analytical method is based solely on voltammetric

Table 2. Aromatic nitro compounds of biological significance

Aromatic nitro compound(s)	Biological significance	Voltammetric technique of analysis	Ref.
Dinitro- and trinitrotoluenes	Explosives	HPLC–ED	Bratin *et al.* (1981)
			Maskarinec *et al.* (1984)
		HPLC–*hv*-ED	Krull *et al.* (1984)
Nitro-substituted polycyclic aromatic hydrocarbons	Diesel engine emission	HPLC–ED	Rappaport *et al.* (1982)
			Jin and Rappaport (1983)
Nitro-containing organophosphorus compounds, e.g. parathion	Pesticides	DPP	Smyth and Osteryoung (1978)
Nitrophenols and nitrochlorobenzenes	Pollutants in natural waters	DPP	Birch and Hart (1979)
Chloroamphenicol, metronidazole, nitrofurantoin	Antibacterial drugs	DPP	Browne (1979)
		HPLC–ED	Sithole and Guy (1986)
Nitrobenzodiazepines	Sedatives	DPP	Clifford and Smyth (1974)
Nifedipine	Coronary vasodilator	DPP	Thoma and Klimek (1980)
		AdSV	Diez-Caballero *et al.* (1989)

determination, limits of detection can be found at ppb levels, e.g. Diez-Caballero *et al.* (1989) quoted a limit of detection of 4.1 ng ml^{-1} in serum for the AdSV determination of nifedipine. These purely voltammetric methods, however, do not possess the necessary selectivity or specificity in subject areas such as metabolic studies or forensic science, respectively. For example, chloramphenicol metabolises in the rat to other aromatic nitro-containing and polarographically unresolvable metabolites such as the oxamic acid, the chloramphenicol base, alcohol and glucuronide and the polarographically unreducible and toxicologically significant chloramphenicol arylamine, where the aromatic nitro group has been converted into an amino group. A simple ion-pair HPLC method has been used for this chloramphenicol metabolic profiling and results in complete separation and identification of all these metabolites (Bories *et al.*, 1983).

From the viewpoint of forensic science, where specificity is of the essence, positive identification of the benzodiazepine flunitrazepam in a blood sample could be taken as an example. Following extraction from the blood sample and differential-pulse polarographic (DPP) analysis, two peaks would be revealed, one corresponding to the nitro group reduction and the other to the azomethine group reduction. This would suggest nitrazepam, clonazepam, flunitrazepam and other structurally similar nitro-containing 1,4-benzodiazepines. Rapid hydrolysis in 0.1 M HCl would indicate flunitrazepam. Absolute confirmation would, however, have to be sought from an analytical technique such as mass spectrometry (Smyth, 1984).

Recent investigations of emissions from diesel engines have focused on the mutagenic and carcinogenic properties of nitro-substituted polynuclear aromatic hydrocarbons (nitro-PAHs). Of the nitro-PAHs which have been tentatively identified in extracts of diesel exhaust particulates, several are potent mutagens in the Ames *Salmonella* bioassay, a short-term test designed to detect chemicals which may be potent carcinogens. These compounds include 1-nitropyrene (1-NP), 2,7-dinitrofluorene (2,7-DNF) and 2- and 3-nitro-9-fluorene (2- and 3-NFO). Other mutagenic nitro-PAHs in diesel exhausts, including 2-nitrofluorene (2-NF) and 4-nitrobiphenyl (4-NB), are known to be carcinogenic in laboratory animals. Numerous other nitro-PAHs have not been tested for mutagenicity or carcinogenicity because of the unavailability of reference compounds. Rappaport *et al.* (1982) and Jin and Rappaport (1983) used HPLC–reductive ED [at − 0.6 V (vs Ag/AgCl)] to measure such nitro-PAHs at picogram and nanogram levels and to identify 1-NP in extracts of diesel exhaust particulates by comparing retention times and the hydrodynamic voltammograms of a reference 1-NP compound (Fig. 3). Preliminary results indicated that some reducible species which might interfere with the analytical method (aldehydes and anhydrides) are not measured at the electrode potential used for nitro-PAHs while others (ketones and quinones) can probably be differentiated from nitro PAHs by their hydrodynamic voltammograms. It is worth noting that a rival analytical method based on reduction of the isolated nitroaromatics from air to the amines with Zn–HCl, followed by derivatisation with hepta-

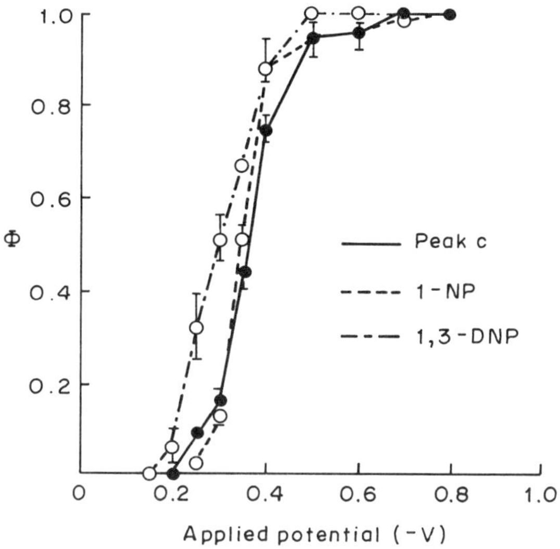

Figure 3. Hydrodynamic voltammograms of peak c [from HPLC of diesel extract No. 1; Rappaport *et al.* (1982)] and reference compounds 1-NP and 1,3-DNP which elute at the same time as peak c under chromatographic conditions—column, 25 cm × 4 mm i.d. Ultrasphere ODS (5 μm spherical particles); mobile phase, 35% propan-1-ol in 0.05 M monochloroacetic acid–sodium acetate buffer (pH 3.8); flow rate, 1 ml min^{-1}; potential of detector, − 0.6 V (vs Ag/AgCl). Reproduced by permission of S.M. Rappaport and Elsevier Science Publishers

fluorobutyric anhydride prior to gas chromatography, is considerably more lengthy (Morita *et al.*, 1982).

Because of the usefulness of the nitro group reduction in voltammetric analysis, there are many instances in the literature where nitration derivatisations have been carried out on phenyl-containing molecules, as shown in Table 3.

An interesting application of derivatisation by nitration is in homogeneous immunoassay where labelled (by nitration, indicated with an asterisk) and unlabelled antigen compete for a limited amount of antibody:

$$\begin{matrix} Ag + \\ & Ab \\ Ag^* + \end{matrix} \begin{matrix} \rightleftharpoons Ab{:}Ag \\ \\ \rightleftharpoons Ab{:}Ag^* \end{matrix}$$

For the bound species Ab:Ag*, the current due to the reduction of the nitro label is decreased by binding with the antibody Ab. This has been demonstrated for the estrogen estriol, which is rendered electroreducible by labelling the phenyl ring with nitro groups to give 2,4-dinitroestriol. A calibration graph of

Table 3. Nitration derivatisation of selected molecules

Molecule(s)	Derivatisation reaction	Notes	Ref.
Phenyl-containing anti-convulsive drugs	Formation of nitro derivative(s)	Application to body fluid analysis using DPP	Brooks *et al.* (1973c)
Amines, amino acids	Formation of derivative with trinitrobenzenesulphonic acid (TNBS)	Pre-column off-line derivatisation prior to HPLC	Jacobs and Kissinger (1982)
Carbonyl compounds	Formation of derivative with 2,4-dinitrophenylhydrazine	Pre-column off-line derivatisation prior to HPLC	Jacobs and Kissinger (1982)
Styrene	Reaction with HNO_3 to give $PhCH(NO)CH_2NO_2$	Styrene determined in polystyrene and its copolymers as the pseudo-nitrosite	Urbansky *et al.* (1977)
Morphine	Treatment with HNO_2 to give 2-nitromorphine rather than the nitroso derivative		Baggesgaard-Rasmussen *et al.* (1945); Lund (1958)
2,4-Dichlorophenoxy-acetic acid (2,4-D)	Nitration of extracted residue from irrigation waters	Limit of detection by DPP is 30 ppb for 2,4-D	Lechien *et al.* (1981)

peak current vs estriol concentration is S-shaped in the concentration range 0.4–5.4 $\mu g\ ml^{-1}$ (Wehmeyer *et al.*, 1982).

1.2 MOLECULES CONTAINING THE NITROSO GROUP

Nitroso-containing molecules, in particular *N*-nitrosamines, are of biological importance as carcinogens and have been detected in a variety of matrices such as synthetic cutting oils, foods and alcoholic beverages. They can be formed *in vivo* by the action of nitrite on secondary and tertiary amines.

N-Nitrosamines generally exhibit an irreversible $4e^-$ reduction in acidic solution to give the corresponding hydrazine and an irreversible reduction in alkaline solution to give N_2O and the precursor amine (Smyth *et al.*, 1975b). The best polarographic waves or peaks for trace analytical purposes are therefore found in acidic media. However, owing to the similarity of reduction potentials, (e.g. in pH 2 buffer the DPP peaks of diphenyl-*N*-nitrosamine, *N*-methyl-*N*-nitrosaniline and *N*-methyl-*N*-nitrosourea are at – 0.60, –0.65 and –0.65 V, respectively (Smyth *et al.*, 1975b). Identification of a particular *N*-nitrosamine is impossible by solely polarographic means. In addition, other electroactive naturally occurring molecules in foods such as furfuraldehyde, cinnamaldehyde and β-ionone can be mistaken for these *N*-nitrosamines, (Walters *et al.*, 1970).

Gel filtration (Smyth *et al.*, 1977b) and HPLC (Samuelsson and Osteryoung, 1981) have been used in attempts to improve the selectivity of the polarographic/voltammetric method. Samuelsson and Osteryoung (1980) used rapid-scan square-wave voltammetry to refine further the selectivity of the detection step. This resulted in a three-dimensional chromatovoltammogram (Fig. 4). Thomas (1986) has written a dissertation on the application of swept-potential ED to the HPLC determination of *N*-nitrosamines and carbamate pesticides. Msimanga and Sturrock (1990) recently applied factor analysis to HPLC with swept-potential ED for a mixture of three *N*-nitrosamines, *N*-nitroso-*N*-methylamine (NMA), *N*-nitrosodi-*N*-propylamine (NDP) and *N*-nitroso-*N*-ethylbutylamine (NBE) all three having very similar capacity factors. NDP and NBE also have identical reduction potentials at – 1.04 V but NMA may be separated from the other two on the potential dimension of a three-dimensional plot, its reduction potential being – 0.84 V. Autocorrelation coefficients, multiple determination coefficients and single-vector uniqueness tests were investigated as predictors of the number of major electroactive components in the mixture, with the latter indicating the presence of three components.

N-Nitrosamines are ideally detected and determined by resort to selective extraction (e.g. clean-up of extract with liquid–liquid partitioning between acetonitrile and *n*-heptane, the *N*-nitrosamines partitioning into the acetonitrile phase and lipophilic constituents of the extract into the *n*-heptane phase), steam distillation (to separate volatile *N*-nitrosamines) and gas chromatography–mass

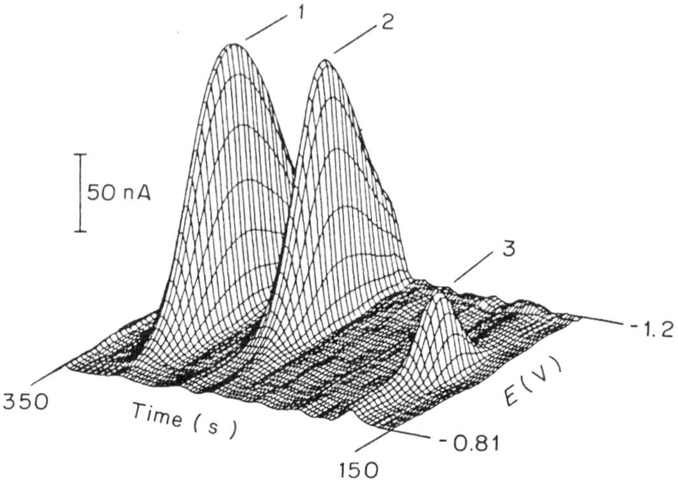

Figure 4. Three-dimensional chromatopolarogram of NDELA and NPRO: (1) NPRO; (2) NDELA; (3) unknown impurity. Conditions: [NPRO] = 1.34×10^{-5} M; [NDELA] = 1.2×10^{-5} M; step height = 10 mV; square-wave amplitude = 25 mV; square-wave frequency = 100 Hz, mobile phase = 1% phosphate buffer (pH 3.5). Background currents have been subtracted. Reprinted with permission from *Analytical Chemistry*, **52**, 2215. Copyright (1980) American Chemical Society

spectrometry (GC–MS) with the single-ion monitor set at $m/z = 30$ for the NO fragment. Concentrations of 1–10 ppb of *N*-nitroso-3-hydroxypyrrolidine can be determined in cured meat products by the application of this approach. Alternatively, detection can be achieved with the thermal energy analyser (TEA) as in HPLC–TEA for non-volatile *N*-nitrosamines.

1.3 OTHER *N*-OXYGENATED MOLECULES

Glennon *et al.* (1989) used reversed-phase HPLC with oxidative ED at + 1.00 V (vs Ag/AgCl) for the determination of medically and industrially important hydroxamic acids. Desferoxamine, $NH_2(CH_2)_5N(OH)CO(CH_2)_2CONH(CH_2)_5N(OH)CO(CH_2)_2CONH(CH_2)_5N(OH)COCH_3$, used in chelation drug therapy, and its Fe(III) complex can be simultaneously determined with limits of detection of 4.1×10^{-8} and 8.4×10^{-7} M, respectively. Hydroxylamines which contain the functional group R′R″NOH and oximes with R‴CH=NOH can be formed by metabolic *N*-oxidation of primary and secondary amines, and Beckett *et al.* (1977) used reductive DPP to monitor such an oxime in the presence of hydroxylamine metabolites and anodic polarography for the determination of a primary hydroxylamine in the presence of a secondary hydroxylamine and the oxime.

Co and Ni can be determined in electronic-grade materials by AdSV of their dimethylglyoxime chelates. In these same materials, Cd, Cu, Pb and Zn are determined by ASV (differential-pulse mode) and Fe by AdSV of its catechol chelate (Metrohm, 1987). The analytical method involves placing one drop of HF on the silicon chip followed by evaporation by warming, leaving a white powdery residue. The latter is dissolved in a drop of concentrated HNO_3 and rinsed into the polarographic vessel with ultra-pure water ($< 0.1\ \mu S\ cm^{-1}$, 20 °C). A blank is prepared using the same reagents. The first voltammetric step is to determine simultaneously Zn(II), Pb(II), Cd(II) and Cu(II) at pH 5 using ASV in the differential-pulse mode. In the second step, Fe(III) is determined in PIPES buffer (pH 7.0–7.2) using AdSV (differential-pulse mode) of the Fe(III)–catechol chelate. Finally, Co(III) and Ni(II) are determined as their dimethylglyoxime chelates in NH_3–NH_4Cl buffer (pH 9.1). In the latter determination, duplicate determination of both Co and Ni is carried out with two standard additions per element. Figure 5 shows the voltammograms corresponding to Ni and Co in the sample. Triethanolamine is used in the preparation of the dimethylgloxime solution and this increases the sensitivity of the Co determination. At the same time, it suppresses possible interferences due to a relatively large Zn content. The Ni peak is found to be somewhat attenuated.

Co and Ni have been determined in sea water at picomolar and nanomolar concentrations, respectively (Donat and Bruland, 1988). Cyclohexane-1,2-dione dioxime (nioxime) chelates of Co(II) and Ni(II) were concentrated by adsorption from 10 ml of sample on to a hanging mercury drop electrode (HMDE), followed by cathodic stripping using the differential-pulse mode. Optimum conditions were ligand concentration 10^{-4} M, HEPES buffer concentration 0.03 M, solution pH 7.6 and accumulation potential − 0.6 V. Replicate analyses

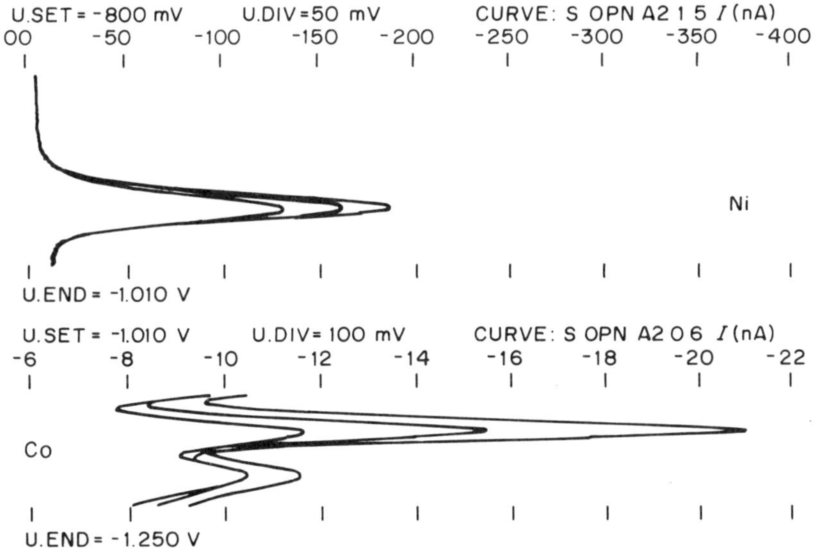

Figure 5. Determination of Co and Ni in electronic-grade materials by AdSV of their dimethylglyoxime chelates. From *Metrohm Application Bulletin*, No. 147, January 1987; reproduced by permission of Metrohm Ltd, Herisau, Switzerland

of sea water reference materials yielded excellent agreement with certified values. The analytical precisions for Co and Ni at coastal and open ocean concentrations were approximately 5% (relative standard deviation). Detection limits for Co and Ni were reported to be 6×10^{-12} and 0.45×10^{-9} M, respectively, after a 15 min accumulation. As an alternative to nioxime, dimethylglyoxime has been used for the determination of Co and Ni in Mediterranean sea water samples (Zhang *et al.*, 1989) and Co in estuarine water samples (Zhang *et al.*, 1988). Nimmo *et al.* (1989) investigated the chemical speciation of dissolved Ni and Co in coastal and estuarine waters and recently (Nimmo, 1991) in rain waters using AdSV of dimethylglyoxime complexes of Ni and Co before and after UV irradiation of the samples. For example, in rain waters, Nimmo (1991) found volume weighted means for Ni concentrations of labile Ni 0.283 $\mu g\ l^{-1}$, total Ni 0.377 $\mu g\ l^{-1}$ and organic Ni 0.093 $\mu g\ l^{-1}$, giving 25% organic Ni, illustrating the atmosphere as an important source of organic Ni. The determination of traces of Ni in NBS fly ash has been automated using a medium exchange flow system and a chemically modified carbon paste electrode containing dimethylgloxime (Thomsen *et al.*, 1988). Preconcentration in the case of Pd has been achieved by adsorption of the dimethylglyoxime chelate on an HMDE in a stirred acetate buffer solution (pH 5.15) containing 2×10^{-4} M ligand and an accumulation potential of − 0.20 V. For a 10 min preconcentration time, the limit of detection is 20 ng l^{-1} (2.1×10^{-10} M) (Wang and Varughese, 1987).

N-Oxides, metabolic products of tertiary amines, are, as with the hydroxylamines, thermally labile when subjected to GC so polarographic methods can present an alternative for their trace determination. Beckett *et al.* (1974) devised such a method for the determination of the *N*- and *S*-oxide metabolites of the major tranquilliser chlorpromazine in biological samples based on pH control, ion-pair formation, selective solvent extraction and cathode-ray polarography. Billedeau *et al.* (1990) recently used HPLC–ED in the analysis of an extract from a 144 h incubation of pyrilamine maleate (35 μmol) with a fungal culture of *C. elegans*. Figure 6 illustrates the resolution of this antihistamine, pyrilamine (**I**), its major biotransformation product, the *N*-oxide, *N*-demethylated pyrilamine (*N*-DMe) and *O*-demethylated pyrilamine (*O*-DMe). Tandem UV detection at 254 nm, as illustrated in Fig. 6, shows less specificity in that the *N*-oxide can be identified by the presence of its electrochemical response at + 0.9 V and the absence of its electrochemical response at + 0.7 V.

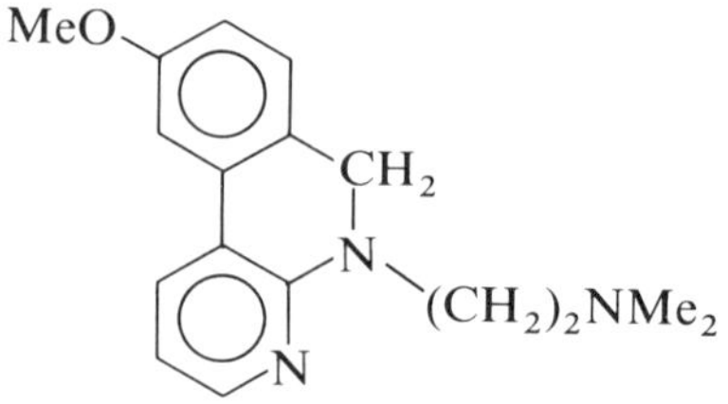

(**I**)

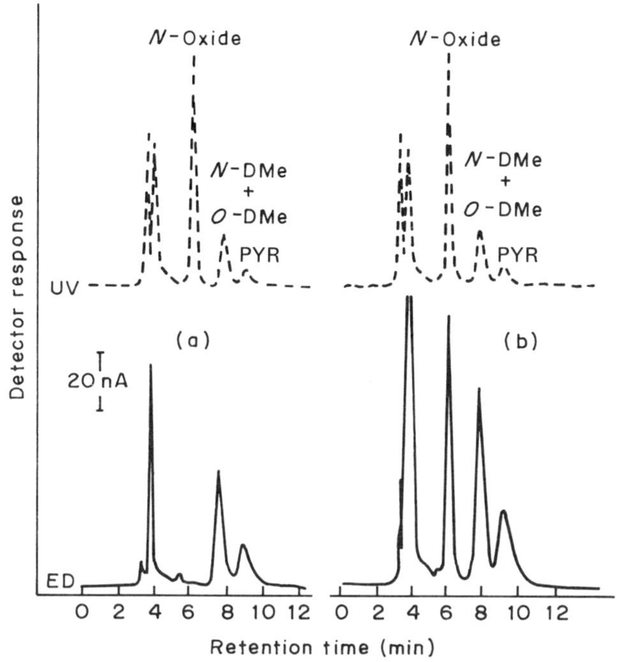

Figure 6. HPLC analysis of a 1:250 dilution of a 144 h microbial transformation of pyrilamine maleate, 35 μmol, by a fungal culture of *C. elegans* using ED at applied potentials of (a) + 0.7 and (b) + 0.9 V with simultaneous UV detection at 254 nm (dashed lines). Peaks: *N*-Oxide = pyrilamine *N*-oxide: *N*-DMe = *N*-demethylated pyrilamine: *O*-DMe = *O*-demethylated pyrilamine; PYR = pyrilamine. Reproduced by permission of Elsevier Science Publishers

Sestakova and Kopanica (1988) used a combination of gel chromatography or solvent extraction with AdSV for the determination of the growth promoter cyadox, a quinoxaline-*N*-dioxide, and its metabolites in the plasma from pig blood. Figure 7 shows that the results obtained by these two techniques are in good agreement with those for radiometric determination of the plasma level after administration (at 80 mg day^{-1} for 7 days) of cyadox labelled with tritium in the benzene ring.

1.4 MOLECULES CONTAINING THE AZOMETHINE GROUP

In the history of polarography, there have been many instances where the technique has been used for the determination of compounds containing the C=N functional group. This has been particularly true for the 1,4-benzodiazepines, a series of drug substances that are widely prescribed as anti-anxiety

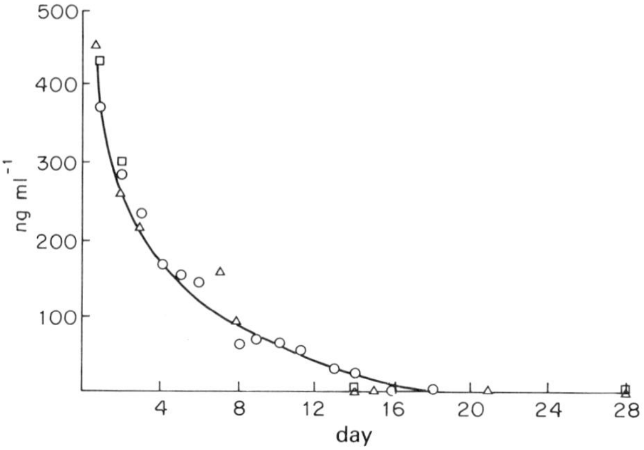

Figure 7. Levels of the cyadox metabolites in the plasma from pigs (ng ml^{-1}) as a function of the number of days elapsed from the termination of cyadox application: △, Adsorptive stripping voltammetry with gel chromatography; □, adsorptive stripping voltammetry with extraction; ○, radiometric measurements. Reproduced with permission from Sestakova and Kopanica, *Talanta*, **35** (1988) 816

agents and sedatives. The electrochemical reduction and oxidation behaviour of 12 important members of this series is summarised in Table 4.

The polarographic reduction peaks correspond to $2e^-$ reductions which saturate the C=N group and, in some cases, to $4e^-$ reductions of the NO_2 group. Several mechanisms of oxidation are apparent for the 1,4-benzodiazepines listed in Table 4 (Smyth and Ivaska, 1985). *N*-Oxidation can take place on

Table 4. E_p values of some 1,4-benzodiazepines in Britton–Robinson buffer (pH 4.0) (E_p values in V vs SCE)

Compound	Polarographic reduction peaks (using DPP)			Voltammetric oxidation peaks (using d.c. mode at GCE[a]
Chlordiazepoxide	−0.37,	−0.73,	−1.17	+1.30
Diazepam		−0.74		NS[b]
Nitrazepam	−0.16,	−0.78		NS
Oxazepam		−0.76		+1.33
Iorazepam		−0.74		+1.38
Prazepam		−0.73		NS
Flurazepam		−0.72		+1.05
Potassium chlorazepate		−0.73		+1.49
Medazepam		−0.82		+0.88, +1.10
Flunitrazepam	−0.16,	−0.73		NS
Clonazepam	−0.15,	−0.73		NS
Bromazepam	−0.40,	−1.12		NS

[a]GCE = glassy carbon electrode.
[b]NS = no signal.

the $—(CH_2)_2N(C_2H_5)_2$ group of flurazepam and the $—NHCH_3$ group of chlordiazepoxide. Radical cations can be formed by removal of $1e^-$ from the N-1 atom for unsubstituted N-1-benzodiazepines such as oxazepam, followed by coupling reactions. When electron density at the N-1 atom is reduced by delocalisation as in 7-NO_2-substituted benzodiazepines such as nitrazepam and/or there is steric shielding of the N-1 atom by a substituent group such as CH_3 as in diazepam, then the molecule lacks electrooxidisability.

Pharmaceutical formulations containing 1,4-benzodiazepines have been assayed by polarographic methods with relative rapidity by mixing the formulation with a suitable solvent in which the active constituent is soluble, dilution of an aliquot of the supernatant with appropriate supporting electrolyte and finally polarographic determination. Errors of the order of 1–2% and similar relative standard deviations are not uncommon. The low limit of detection of modern polarographic methods such as DPP and SWV is found particularly useful in single-tablet assays in quality control. For example. Yarnitzky and Smyth (1991) used rapid-scan SWV with an automated cell for the single-tablet assay of chlordiazepoxide. These polarographic and voltammetric methods can also be used as a rapid means of stability testing formulations, provided that the degradation product(s) are electroactive at different potentials or, alternatively, electroinactive. In the case of oxazepam, the principal degradation product on shelf storage is the quinazoline carboxaldehyde, which will reduce ca 200 mV more positively than the parent compound, enabling both compounds to be determined simultaneously (Goldsmith *et al.*, 1973). Such a polarographic method of formulation analysis is obviously superior to the non-selective official method of assay of the US Pharmacopia employing ethanol extraction followed by UV spectrophotometry or GC in which oxazepam degrades on-column (Sadee and Van der Kleijn, 1971). However, the reversed-phase HPLC method of Reif and Deangelis (1983) has superior selectivity in that a selection of impurities and degradation products can be fully or partially resolved from the parent compound in a single chromatographic run (Table 5).

Table 5. Retention times of oxazepam and related impurities and degradation products using reversed-phase HPLC–UV detection at 254 nm. C_{18} column, mobile phase methanol–water–acetic acid (60 + 40 + 1), flow rate 2 ml min^{-1})

Compound	HPLC retention time (min)
Oxazepam	5.6
Quinazoline carboxaldehyde	11.8
Dione	3.6
Carboxylic acid	4.5
Quinazolinone	8.2
Glyoxanilide	11.2
Alcohol	12.4
Benzophenone	18.4

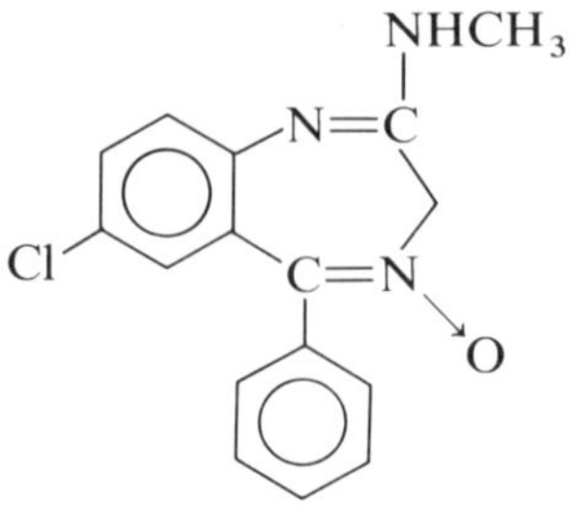

(**II**) Chlordiazepoxide

The polarographic determination of chloradiazepoxide (**II**) in biological fluids employ either dilution of the sample with the supporting electrolyte and measurement (Jacobsen and Jacobsen, 1971) or selective extraction from an alkalinised sample followed by measurement (Cimbura and Gupta, 1965; Hackman *et al.*, 1974). The procedure of Jacobsen and Jacobsen (1971) has a limit of detection (LOD) of 13 μg ml^{-1} of chlordiazepoxide in plasma owing to masking of the polarographic wave by protein adsorbed on the surface of the mercury drop. The method is only suitable as a rapid toxicological assay of the drug in serum due to its high LOD. The assay of Cimbura and Gupta (1965) was applied to the determination of chlordiazepoxide in plasma, urine or gastric contents and is also only suitable for toxicological analysis. The DPP procedure

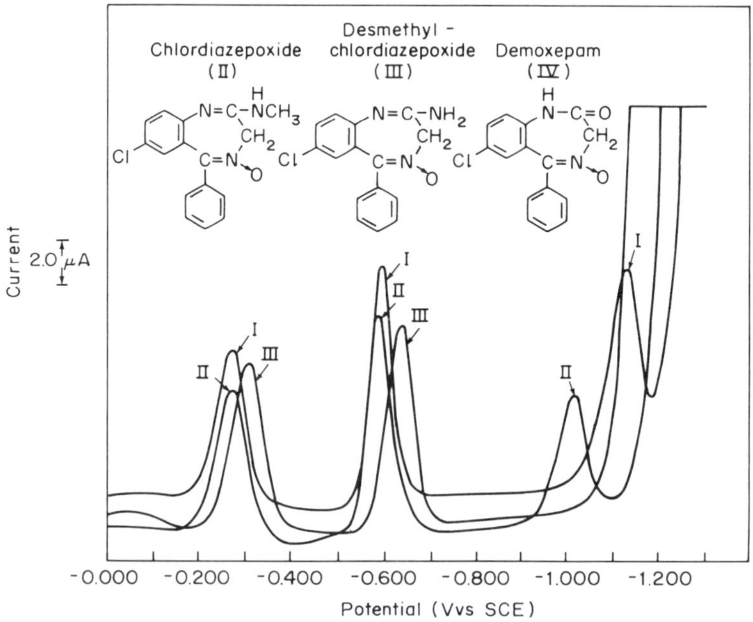

Figure 8. Differential-pulse polarography of chlorodiazepoxide and its metabolites in 0.05 M H_2SO_4. Reprinted with permission from *Analytical Chemistry*, **46**, 1075. Copyright (1974) American Chemical Society

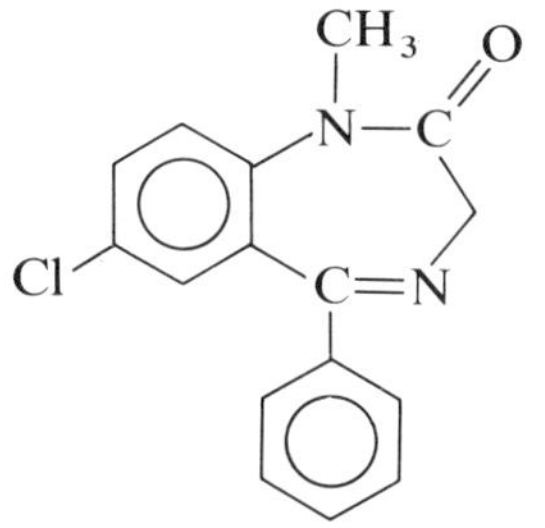

(V) Diazepam

of Hackman *et al.* (1974) is capable of determining the drug and its *N*-desmethyl (**III**) and lactam (**IV**) metabolites following therapeutic administration. Figure 8 shows the DPP behaviour of these substances in 0.05 M H_2SO_4. The LOD of this latter assay is 0.05–0.1 μg ml^{-1} of each compound in serum using a 2 ml sample. The assay is rapid and selectivity is improved by the thin-layer chromatographic step. It was used to measure levels of parent drug and metabolites following a single 30 mg i.v. and oral dose and following chronic oral administration. A comparison of the data obtained using this assay and the standard spectrofluorimetric assay (Table 6) showed a good correlation between the methods. In addition, the DPP assay measured the intact compounds, having the advantage over the spectrofluorometric assay that the photochemical conversion of the compounds to fluorescent derivatives was eliminated.

Toxicological analysis for diazepam (**V**) in blood, urine or gastric contents

Table 6. Plasma levels (μg ml^{-1}) of chlordiazepoxide and its metabolites following chronic administration[a] of a single daily 30 mg oral dose. Reprinted with permission from Hackman *et al.*, *Anal. Chem.*, **46**, 1075. Copyright (1974) American Chemical Society.

Study interval		Chlordiazepoxide (**II**)		*N*-Desmethyl-chlordiazepoxide (**III**)		Demoxepam (**IV**)	
Day	Hour	DPP[b]	Fluor.[c]	DPP	Fluor.	DPP	Fluor.
22	1	1.65	1.42	0.41	0.33	0.43	0.42
22	2	1.63	1.46	0.41	0.29	0.46	0.42
26	1	1.45	1.36	0.40	0.32	0.50	0.48
26	2	1.28	1.18	0.33	0.26	0.53	0.49
30	1	1.57	1.48	0.35	0.33	0.49	0.48
30	2	1.58	1.35	0.37	0.32	0.50	0.55
34	1	1.43	1.50	0.31	0.26	0.43	0.52
34	2	1.33	1.35	0.41	0.27	0.46	0.47
36	1	1.34	1.41	0.36	0.28	0.40	0.48
36	2	1.23	1.23	0.37	0.27	0.41	0.62

[a]The subject received 10 mg oral doses three times a day for 21 days, and a single 30 mg oral dose once a day thereafter.
[b]Differential-pulse polarographic procedure.
[c]Spectrofluorometric procedure.

(Cimbura and Gupta, 1965) and in blood (Fidelus *et al.*, 1972) have been reported. The determination of therapeutic levels of the drug in plasma by linear-sweep voltammetry (Berry, 1971) and in serum by DPP (Jacobsen *et al.*, 1973) have also been reported. These assays have sensitivities of 0.02–0.03 μg ml^{-1} of diazepam; however, they are considered 'total' assays since they also measure levels of the *N*-desmethyl metabolite which will be co-extracted using the described procedures. Brooks *et al.* (1973b) developed a DPP method incorporating a thin-layer chromatographic separation for the determination of both diazepam and *N*-desmethyl diazepam extracted from blood. The method differed from the other reported assays in that diethyl ether replaced light petroleum or benzene as the extraction solvent. The more polar solvent was found necessary to extract quantitatively both the parent drug and its metabolite in order to measure both compounds with high sensitivity. Brooks and de Silva (1975) have demonstrated that diazepam and its desmethyl metabolite can be measured with an LOD of 0.05–0.1 μg ml^{-1} using a 2 ml plasma sample. Thus the assay is capable of determining the levels of diazepam arising after 10 mg single dose administration, and of determining both the parent drug and the metabolite with the necessary selectivity for therapeutic or toxicological examination.

The urinary metabolites of diazepam in man have been evaluated by Dugal *et al.* (1973) using rapid-scan polarography and by Brooks and de Silva (1975) using DPP. The former procedure requires conversion of the parent drug and metabolites into their respective benzophenones prior to polarography. The DPP procedure employed enzymatic deconjugation of the oxazepam, *N*-desmethyldiazepam and 3-hydroxydiazepam metabolites, selective extraction and thin-layer chromatography prior to polarographic analysis. The assay measured all three compounds with an LOD of approximately 0.25–0.5 μg ml^{-1} in urine and monitored the urinary levels following therapeutic doses of diazepam.

de Silva *et al.* (1974a) described a DPP assay to measure levels of lorazepam (**VI**) glucuronide in the urine following a 4 mg dose of 7-chloro-1,3-dihydro-5-(2′-chloro-phenyl)-2*H*-1,4-benzodiazepin-2-one. The assay required deconjugation of lorazepam at pH 5.3, followed by extraction of the drug into diethyl ether

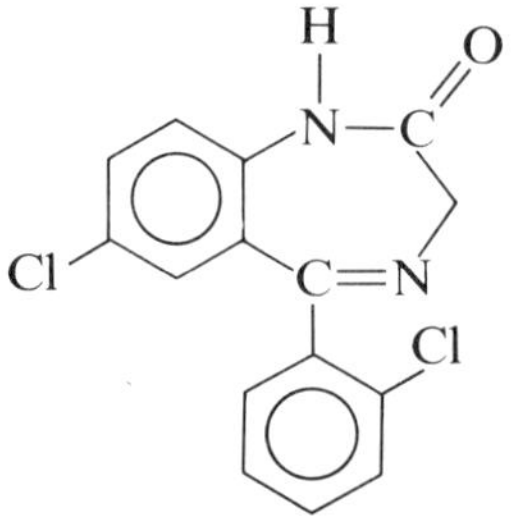

(**VI**) Lorazepam

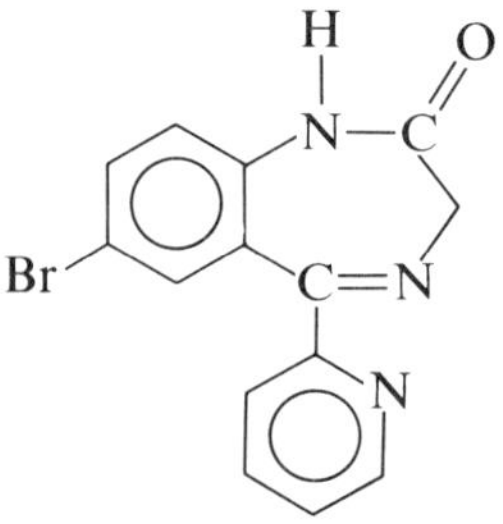

(**VII**) Bromazepam

after adjustment of the pH to 9.0. The residue of the ether extract was dissolved in 0.1 M HCl containing 5% methanol, and analysed by DPP. The assay was rapid, and had an LOD of 0.05 μg ml^{-1} using a 5 ml sample. The assay possessed a lower LOD than the standard Bratton–Marshall spectrophotometric assay.

de Silva *et al.* (1974b) determined the urinary excretion products of bromazepam (**VII**) in man using DPP following the administration of a single 12 mg oral dose. The assay involved selective extraction of bromazepam and the 2-amino-5-bromobenzoylpyridine metabolite from the deconjugated metabolites 3-hydroxybromazepam and 2-amino-3-hydroxy-5-bromobenzoylpyridine into separate diethyl ether fractions. The residues of the respective diethyl ether extracts were dissolved in 1 M phosphate buffer (pH 5.4) and analysed by DPP, which yielded two distinct, well resolved reduction peaks for the 4,5-azomethine functional group of the benzodiazepine-2-one and for the carbonyl functional group of the benzoylpyridine in each fraction. The overall recoveries of the compounds in the conjugated and unconjugated fractions were 80 ± 5% and 45 ± 5%, respectively, with an LOD for the four compounds of 50–100 ng per 5 ml of urine.

Brooks and Hackman (1975) determined bromazepam in blood with an LOD of 10–20 ng ml^{-1} using a 0.5 ml micro-cell for DPP analysis. The assay employed selective extraction of bromazepam from alkalinized blood into benzene–methylene chloride (90:10). The residue of this extract was dissolved in phosphate buffer (pH 7.0) for analysis. The DPP and GC (electron-capture detection) assays measure blood levels in man following chronic single daily oral administrations of 3 mg of bromazepam. The two assays yielded comparable results (see Table 7). The chromatographic and DPP methods have comparable LODs, although the chromatographic method is more selective. Another study

Table 7. Blood levels of bromazepam following chronic administration of a single daily 3 mg oral dose. Reprinted with permission from Brooks and Hackman, *Anal. Chem.*, **47**, 2059. Copyright (1975) American Chemical Society

	Concentration (ng ml^{-1})	
Day	DPP[a]	GC-ECD[b]
1	26	27
4	48	53
9	69	92
10	71	77
12	123	125
15	119	105

[a]Differential-pulse polarographic assay.
[b]Gas chromatography with electron-capture detection.

described the determination of bromazepam in blood by linear-sweep voltammetry with an LOD of 0.05 μg ml^{-1} (Senguen and Oelschlager, 1975).

Smyth *et al.* (1977a) made a spectral and polarographic study of the acid–base and complexing behaviour of bromazepam. An indirect analytical method, based on the complexation of bromazepam with Cu(II) ions, was investigated by these authors (Fig. 9).

Nitrazepam, a hypnotic and sedative drug, is similar in structure to diazepam and oxazepam. The compound is identical with *N*-desmethyldiazepam with the exception that the 7-chloro group has been replaced by a 7-nitro group in the molecule. Senkowski *et al.* (1964) were the first to report the polarography of this compound. The authors noted two polarographic waves in 20% methanol–0.1 M HCl for the reduction of the 7-nitro group and the reduction of the 4,5-azomethine bond, respectively. Further studies (Oelschlager *et al.* 1966) demonstrated that the first polarographic wave was due to the four-electron reduction of the 7-nitro substituent to the hydroxylamine and the second to the simultaneous four-electron reduction of the 4,5-azomethine and the intermediate hydroxylamine to the amine. Under alkaline conditions, four electrons were involved in the reductions of the 7-nitro group and only two electrons for the reduction of the 4,5-azomethine. Oelschlager *et al.*, (1969), Halvorsen and Jacobsen (1972) and Clifford and Smyth (1973) also studied in detail the reduction mechanism of nitrazepam in acidic, neutral and basic media and proposed several polarographic reduction intermediates. The reduction mechanism is shown in Fig. 10 and a cyclic voltammogram in Fig. 11.

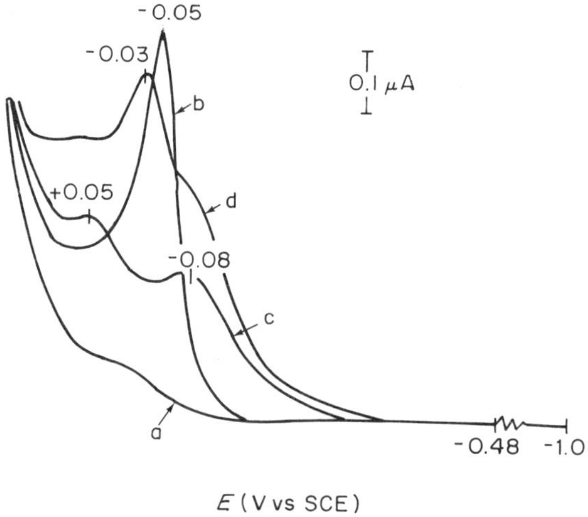

Figure 9. Effect of bromazepam on stripping current of Cu^{II} ions: (a) blank Britton–Robinson buffer (pH 9.0); (b) 1×10^{-5} M Cu^{II}; (c) 1×10^{-5} M $Cu^{II} + 1 \times 10^{-5}$ M bromazepam: (d) 2×10^{-5} M $Cu^{II} + 1 \times 10^{-5}$ M bromazepam. Reproduced by permission of Elsevier Science Publishers from Smyth *et al.* (1977a)

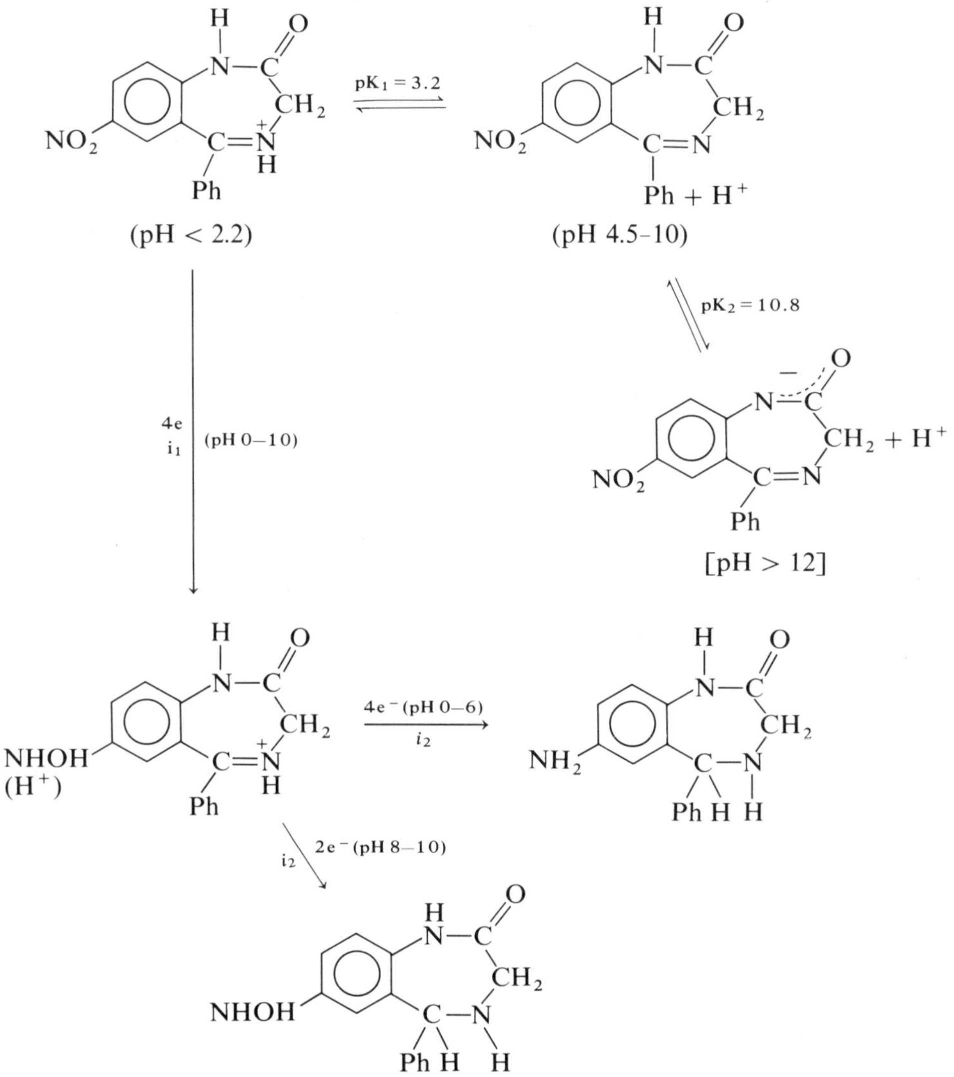

Figure 10. Polarographic reduction mechanism and acid–base equilibrium of nitrazepam. Reproduced from Clifford and Smyth (1973) by permission of Springer-Verlag.

Oelschlager *et al.* (1967) also described a method for the determination of nitrazepam and its 7-amino and 7-acetamido metabolites by incorporating a thin-layer chromatographic separation into the analysis. All three compounds in the range of 30–90 μg were separated, eluted from the thin-layer adsorbent and determined by d.c. polarography with a precision of 4 %. The polarography of the two metabolites showed the typical reduction of the 4,5-azomethine bond. These metabolites were examined in Britton–Robinson buffers in the concentration range of 5×10^{-5}–10^{-3} M in supporting electrolytes of pH 2–10.

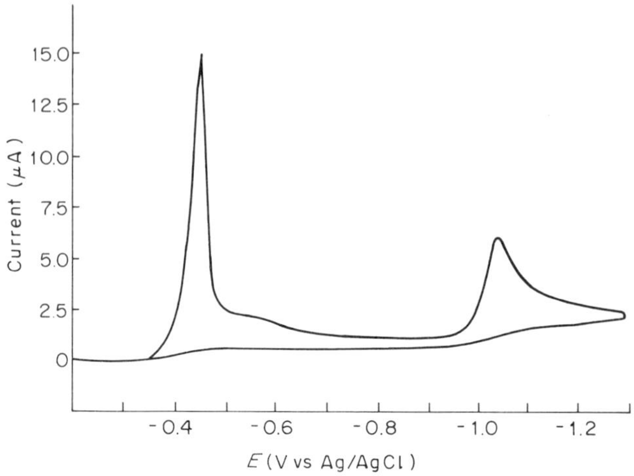

Figure 11. Cyclic voltammogram of 10^{-4} M nitrazepam in phosphate buffer (pH. 6.9). Scan rate 0.1 V s^{-1}. Reproduced by permission of Elsevier Science Publishers from Halvorsen and Jacobsen (1972)

Flurazepam is a hypnotic drug which has been introduced for the treatment of insomnia. A urinary assay developed by de Silva *et al.* (1974c) can determine the parent drug and its major metabolites. The assay involved the determination of the major unconjugated metabolites, *N*-monodesalkyl- and *N*-didesalkylflurazepam, and the conjugated metabolite, *N*-1-hydroxyethylflurazepam, and employed a combination of selective extraction, thin-layer chromatographic separation and DPP. The supporting electrolyte was pH 4 phosphate buffer, and the compounds all exhibited polarographic E_p values of -0.80 ± 0.03 V (vs SCE). The assay was applied to the measurement of urinary levels of the metabolites of flurazepam following a single 90 mg dose (Table 8). An assay in which 0.1 ml of urine was diluted with 1 M phosphate buffer (pH 7.0) and measured ($E_p = -0.955$ V vs SCE) directly was also presented. This assay with an LOD of 5 μg ml^{-1} was designed to be used as a rapid means of measuring 'total' benzodiazepines to confirm ingestion of flurazepam. The rate of hydrolysis of flurazepam was monitored by simultaneously measuring the height of the reduction peaks for the 4,5-azomethine group and for the carbonyl group of its benzophenone derivative at -0.595 and -0.715 V (vs SCE), respectively, and indicated a reaction half-life of 44.5 min. This reaction was found to be completely reversible by adjustment of the pH to 7.0.

Two other reports have dealt with the measurement of flurazepam and its metabolites in plasma (Clifford *et al.*, 1974) and in blood (Brooks and Hackman, 1975) by DPP. The assays have similar limits of detection (10–20 ng ml^{-1}) and could be used to measure levels of *N*-desalkylflurazepam following chronic administration or an overdose of flurazepam. Levels of the parent drug and *N*-1-hydroxyethyl metabolite are below the limits of detection.

Table 8. Urinary excretion data on the major metabolites of flurazepam in humans who had received single oral 90 mg doses of Dalmane. Reproduced by permission of Academic Press

Subject	Excretion period (h)	Form excreted	Compound measured[a]	Concentration ($\mu g\ ml^{-1}$)	Milligram equivalent as flurazepam	Percentage of dose excreted	Direct assay for total benzodiazepines (%)
1	0–24	Free	AI	0.17	0.35	0.49	
			AII	1.14	2.75	3.63	
		Bound	B	9.30	22.5	29.70	
						Total: 33.82	42.2
2	0–24	Free	AI	1.83	2.26	2.98	
			AII	6.57	8.78	11.6	
		Bound	B	18.1	24.3	32.10	
						Total: 46.68	50.2
3	0–24	Free	AI	2.14	1.73	2.28	
			AII	11.4	9.95	13.1	
		Bound	B	25.8	22.6	29.80	
						Total: 45.18	43.8

[a]AI = monodesethylflurazepam; AII = didesethylflurazepam; B = *N*-1-hydroxyethylflurazepam.

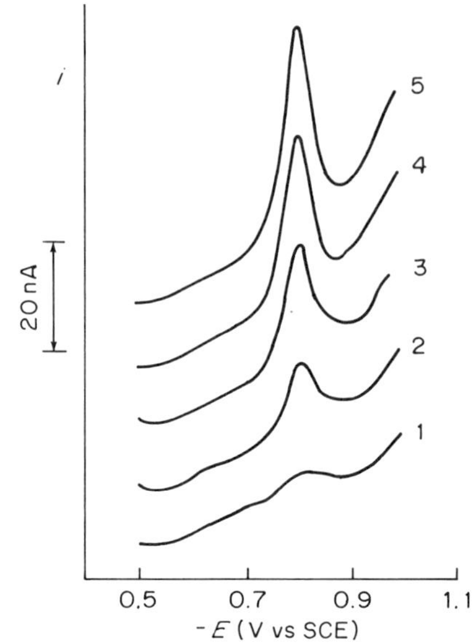

Figure 12. Influence of preceding adsorptive accumulation on the peak height for 5×10^{-8} M diazepam in acetate buffer (pH 4.6). Accumulation time: (1) 0; (2) 30; (3) 60; (4) 120; (5) 180 s. Conditions: DPV, $E_{acc} - 0.5$ V. Reproduced by permission of Elsevier Science Publishers

Groves (1976) has made a detailed study of the acid–base and polarographic behaviour of flurazepam and its metabolites and his comprehensive work should be consulted for further reference.

1,4-Benzodiazepines can be determined by AdSV with limits of detection in the range 10^{-9}–10^{-10} M. The molecules are adsorbed at potentials more positive than the reduction potentials and the determination step made by, for example, DPP at a static mercury drop electrode (SMDE) or hanging mercury drop electrode (HMDE). This is illustrated in Figs 12 and 13 for diazepam (Kalvoda, 1984). Some of the results published to date are collected in Table 9.

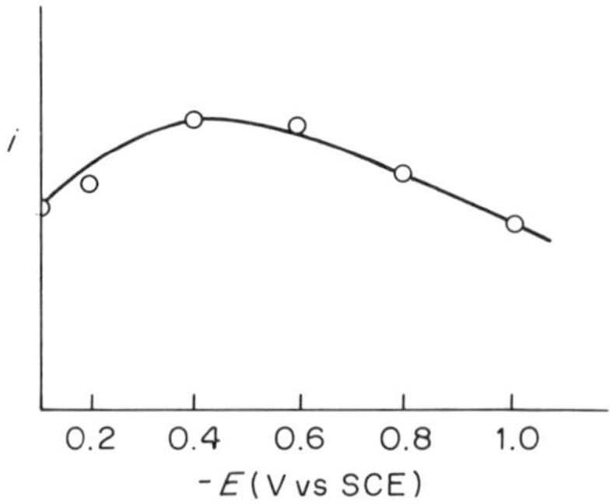

Figure 13. Dependence of the peak height of 2×10^{-6} M diazepam in acetate buffer. (pH 4.6) on E_{acc} for a 60 s accumulation time. Reproduced by permission of Elsevier Science Publishers

Table 9. AdSV data for some 1,4-benzodiazepines

1,4-Benzodiazepine	Supporting electrolyte[a]	Linear calibration range (M)	Limit of detection (M)	Ref.
Diazepam, Nitrazepam	Acetate buffer	—	5×10^{-9}	Kalvoda (1984)
Camzaepam	BR[a] buffer (pH 5)	—	3.4×10^{-10}	Hernandez *et al.* (1987)
Bromazepam			3.5×10^{-9}	
Flurazepam	0.1 M acetate buffer	—	10^{-8}	Smyth and Yarnitzky (1988)
Prazepam	0.2 M acetate buffer (pH 5)	$(2–9) \times 10^{-8}$	4×10^{-9}	Hernandez *et al.* (1986)
Potassium chlorazepate	0.04 M acetate buffer (pH 5)	$(1–9) \times 10^{-8}$	3×10^{-9}	
Medazepam	0.04 M BR buffer (pH 5)	$(4–9) \times 10^{-8}$	2×10^{-9}	
Loprazolam mesilate	0.4 M NH_4Cl (pH 4)	—	2.5×10^{-10}	Arcos *et al.* (1990)
Midazolam	0.1 M acetate buffer (pH 4.8)	—	10^{-10}	Ribes and Osteryoung (1990)
Bentazepam	0.08 M phosphate buffer (pH 8)	—	2.7 ng ml^{-1}[b]	Hernandez *et al.* (1990)
Metaclazepam	BR buffer (pH 9.1)		0.55 ng ml^{-1} (urine)	Gonzalez and Hernandez (1991)
Alprazolam	0.1 M phosphate buffer (pH 7)	1–30 ng ml^{-1}	0.3 ng ml^{-1}[b]	Habeeb *et al.* (1989)
Chlordiazepoxide	BR buffer (pH 6.8)	—	6.9×10^{-8}[c]	Lorenzo and Hernandez (1987)
Temazepam	BR buffer (pH 3.0)	—	17 ng ml^{-1}[b]	Zapardiel *et al.* (1988)
Lorazepam	0.02 M BR buffer (pH 2.0)	—	15 ng ml^{-1}[b]	Zapardiel *et al.* (1990)

[a]BR = Britton–Robinson.
[b]Applies to urine extract following solvent extraction.
[c]Applied to serum extract following Sep-Pak C_{18} cartridge separation.

Zapardiel *et al.* (1991) carried out stripping voltammetric studies on the interaction between the benzodiazepines camazepam, temazepam and oxazepam and human serum albumin. Three-dimensional peak current–accumulation time–albumin concentration plots are given in Fig. 14, illustrating protein binding of the particular drugs. Furthermore, adsorption of these benzodiazepines at a mercury electrode is favoured by the presence at C-3 of bulky groups such as the ester group of camazepam and also by a methyl group at N-1. Table 10 summarises the equations for the calibration graphs obtained for albumin concentrations in the range 0–10 μg ml^{-1}. Zapardiel *et al.* (1991) suggested that a comparison of the slopes for the different drugs could help establish the degree of interaction between the drugs and serum albumin. In this same paper, the authors developed an AdSV method for oxazepam at the HMDE in 0.008 M Britton–Robinson buffer (pH 2) with $-$ 0.50 V accumulation potential. The LOD is 3.6×10^{-10} M for 30 s accumulation with a relative standard deviation of $<2.8\%$ at concentrations in the range 2.8×10^{-8}–4.10^{-7} M (80 s accumulation).

Selavka *et al.* (1985) investigated the use of photolytic derivatisation for improved HPLC–ED of pharmaceuticals in biological fluids. In particular, they compared HPLC–*hv*–ED with GC–MS for the determination of the 1,4-benzodiazepine chlordiazepoxide and its metabolite norchlordiazepoxide in urine and found good agreement between the two methods for the samples analysed; e.g. GC–MS gives a value of 300 ppb and HPLC–*hv*–ED 311 ppb for the chlordiazepoxide content of sample 70-2. Figure 15 shows the two analytes of interest clearly resolved from matrix components and their retention times agree with those observed for the injection of a mixed standard, shown in the centre of Fig. 15. A third injection, containing the sample 70-2 extract solution, was made after the lamp had been extinguished and demonstrates the necessity for photolytic derivatisation prior to ED at the potentials indicated in Fig. 15.

Stewart and Carter (1986) applied HPLC–ED using a glassy carbon indicator electrode held at + 1.30 V (vs SCE) to determine the analgesic chlorzoxazone (**VIII**) and its hydroxy metabolite with a limit of detection 30 times lower than by HPLC–UV. The mechanism of electrooxidation is likely to involve the nitrogen atom in the heterocyclic ring.

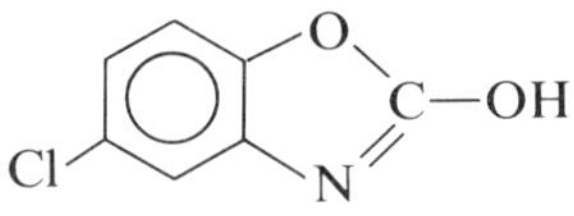

(VIII)

AdSV has recently been applied to the determination of Pt in blood (Nygren *et al.*, 1990) for use in the estimation of natural Pt levels, for monitoring patients treated with Pt-containing cytotoxic drugs and for monitoring occupational exposure to these drugs and other Pt-containing compounds. Blood samples were dry ashed in a muffle furnace and Pt was determined by adsorption of the C=N-containing Pt formazone complex, which results in catalytic evolution of

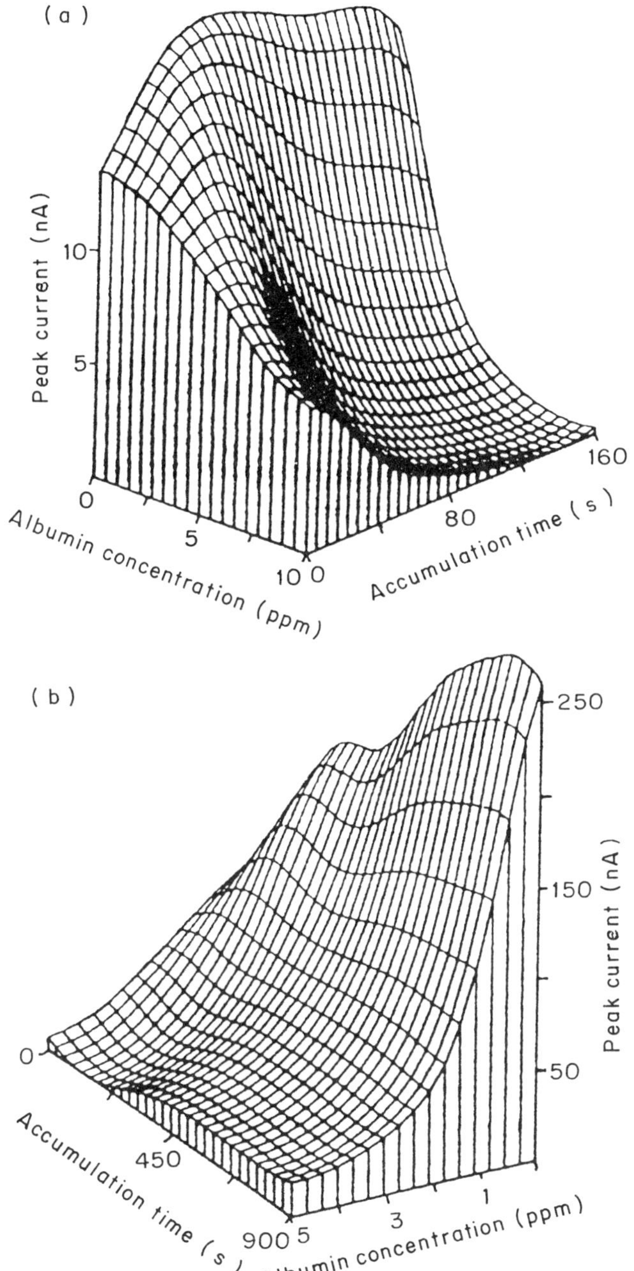

Figure 14. Peak current vs accumulation time for (a) 4.90×10^{-8} M temazepam and (b) 3.14×10^{-8} M camazepam with concentrations of albumin between 0 and 10 μg ml^{-1}. Accumulation potential: (a) − 0.50 and (b) − 0.60 V. Electrolytes: (a) 0.01 M Britton–Robinson buffer (pH 3.0) and (b) 0.04 M acetate buffer (pH 5.0). Scan rate, 20 mV s^{-1}; pulse amplitude, 70 mV. Reproduced by permission of Elsevier Science Publishers

Table 10. Calibration graphs for the drugs with albumin concentrations between 0 and 10 μg ml^{-1}. Reproduced by permission of Elsevier Science Publishers from *Anal. Chim. Acta*, **244** (1991) 49

Albumin	Camazepam		Temazepam		Oxazepam	
concentration (μg ml^{-1})	Calibration equation	Correlation coefficient[a]	Calibration equation	Correlation coefficient[a]	Calibration equation	Correlation coefficient[a]
0.0	$I = 0.95 + (6.94 \times 10^8 C)$	0.9994	$I = -0.23 + (2.46 \times 10^8 C)$	0.9998	$I = 1.07 + (2.03 \times 10^8 C)$	0.9991
2.0	$I = -3.42 + (7.76 \times 10^8 C)$	0.9985	$I = -1.17 + (2.54 \times 10^8 C)$	0.9996	$I = -0.32 + (1.80 \times 10^8 C)$	0.9996
4.0	$I = -1.78 + (6.69 \times 10^8 C)$	0.9972	$I = -0.56 + (9.65 \times 10^7 C)$	0.9992	$I = -0.86 + (1.46 \times 10^8 C)$	0.9994
6.0	$I = -4.59 + (1.90 \times 10^8 C)$	0.9894	$I = -0.49 + (5.61 \times 10^7 C)$	0.9980	$I = -1.03 + (7.30 \times 10^7 C)$	0.9998
8.0	$I = -1.04 + (2.88 \times 10^7 C)$	0.9958	$I = -0.66 + (2.03 \times 10^7 C)$	0.9969	$I = 0.29 + (5.30 \times 10^7 C)$	0.9994
9.0	$I = -0.39 + (1.58 \times 10^7 C)$	0.9924			$I = -0.15 + (4.30 \times 10^7 C)$	0.9989
10.0					$I = -0.71 + (4.30 \times 10^7 C)$	0.9993

[a]10 points.

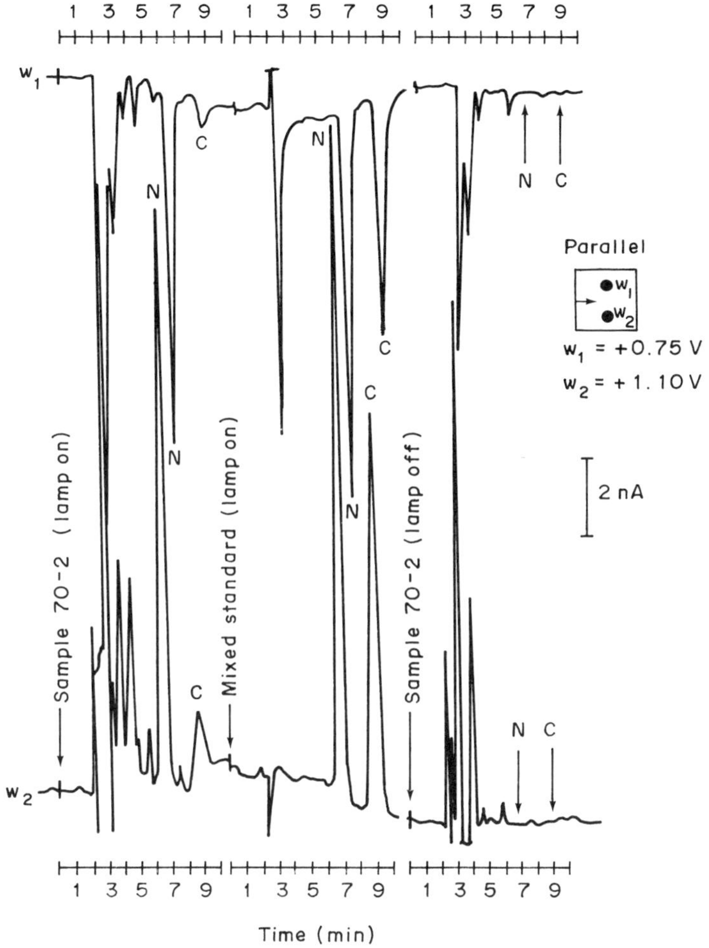

Figure 15. HPLC-*hν*-ED for the assay for chlordiazepoxide and norchlordiazepoxide in human urine. Column: Alltech C_{18}, 10 μm, 250 mm × 4.6 mm i.d. Eluent: methanol-water (45 + 55) containing 7 ml l^{-1} of butylamine; pH, 3.1 (H_2SO_4); flow rate, 1.9 ml min^{-1}. N = norchlordiazepoxide; C = chlordiazepoxide; mixed standard contains both compounds at the 0.5 ppm level. From Selavka *et al.* (1985). Reproduced from the *Journal of Chromatographic Science* by permission of Preston Publication, A Division of Preston Industries, Inc.

hydrogen following reduction of available protons (Wang *et al.*, 1987b) (Fig. 16). This effect has also been observed in the AdSV of the antibiotic cephalothin, (Peled *et al.*, 1987) at the SMDE. The detection limit for a 100 μl sample of blood is 0.017 ppb, with a recovery 94 % and a relative standard deviation 7 % at a Pt level of 1 ppb. The results were verified by inductively coupled plasma mass spectrometry (ICP-MS) with blood prepared by wet ashing and using gold as an internal standard (Table 11).

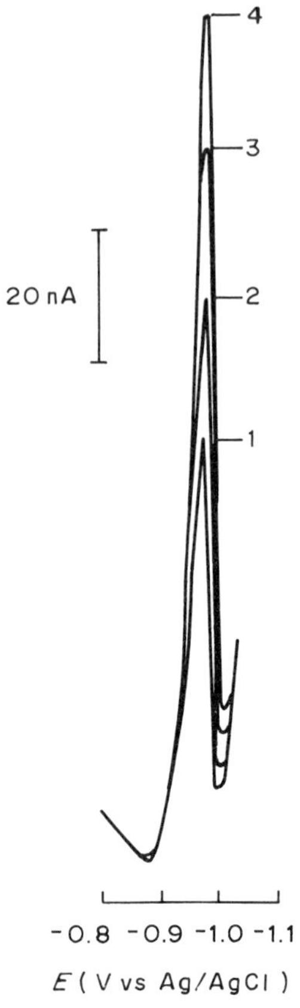

Figure 16. Determination by AdSV of platinum in a 100 μl sample of blood after dry ashing. The scans of the sample solution (curve 1) and three subsequent standard additions of platinum (20 pg) were obtained after 120 s of deposition in the presence of 0.026 % formaldehyde, 0.0032 % hydrazine and 0.7 M HCl (curve 2, curve 1 + 20 pg of Pt; curve 3, curve 1 + 40 pg of Pt; curve 4, curve 1 + 60 pg of Pt). Reprinted with permission from *Analytical Chemistry*, **62**, 1637. Copyright (1990) American Chemical Society

When the C=N group is bonded to another nitrogen atom to form the C=N—N moiety, the electrode process usually involves simultaneous saturation of the azomethine moiety and reductive splitting of the N—N bond. This occurs for several compounds of pharmaceutical interest, e.g. benzhydrylpiperazines (Smyth *et al.*, 1976a, b) and nitrofurantoin (Burmicz *et al.*, 1976). DPP has

Table 11. Determination of blood Pt levels by ICP-MS after wet ashing and AdSV after dry ashing

	Pt concentration (ppb)	
Sample	ICP-MS	AdSV
A	1.71 ± 0.04	1.66 ± 0.11
B	0.61 ± 0.03	0.58 ± 0.03
C	0.47 ± 0.01	0.51 ± 0.05
D	0.85 ± 0.02	0.74 ± 0.06
E	2.33 ± 0.03	2.24 ± 0.16

been used to determine the concentration of one of these benzhydrylpiperazines, SC-13504, in the plasma of baboons following therapeutic administration and a relationship has been obtained between the plasma levels of SC-13504 and its anticonvulsant activity in this animal species (Meldrum *et al.*, 1976). Smyth *et al.* (1977c) also made comparisons between the plasma levels of SC-13504 measured by DPP and those measured by radiochemical and GC methods with a good correlation being obtained.

1.5 MOLECULES CONTAINING THE AZO GROUP

Compounds containing the azo group are biologically significant in that they are commonly used as dyestuffs and food additives. The azobenzene–hydrazobenzene system gives rise to a $2e^-$ reversible process. In solution of a strong acid, hydrazobenzene can be further reduced to aniline in an irreversible process at more negative potentials. Florence (1974) investigated in detail the mechanism of the polarographic reduction of a variety of substituted azobenzenes over a wide range of pH. He found that substitution of a nitro group stabilised the hydrazo intermediate formed initially, resulting in processes consuming $2e^-$ over a wide pH range. Substitution of hydroxyl group in the *meta* position also resulted in identical $2e^-$ processes, e.g. 3,3′-dihydroxyazobenzene. However, substitution of hydroxy groups in the *ortho* and *para* positions gave rise to $4e^-$ processes in acidic media, which fell off to a value approaching $2e^-$ in neutral and alkaline media. The phenylazonaphthol dyes 1-phenylazo-2-naphthol and 1-(2-hydroxyphenylazo)-2-naphthol gave rise to $4e^-$ processes in the pH range 2–10. The reversible reduction to the hydrazo intermediate and irreversible reduction to the resulting amines was confirmed by cyclic voltammetry. As with aromatic nitro compounds, electrochemical studies of this kind can be important in assessing the toxicity of a new compound (Carruthers, 1976). In this context, Smyth and Hassanzadeh (1976) used DPP to monitor selected azo dyes in an aquarium environment and found that Acid Red 73 was the most toxic to daphnae and plant life. This textile dye contains the above-mentioned 1-phenylazo-2-naphthol group and was the easiest of the four azo dyes investigated to reduce polarographically (−0.18 V in formate buffer). Its toxicity could therefore be connected with a readily accomplished metabolic reduction of the —N=N— group in Acid Red 73 to yield toxic aromatic amines.

Fogg *et al.* (1986) examined sixteen food and two cosmetic synthetic colouring matters by AdSV using the differential-pulse mode and found that when a 2 min accumulation time was used, the increase in sensitivity over DPP at the DME was between 8- and 100-fold, depending on the colouring matter being determined. Using longer accumulation times, 10^{-10} M concentrations of some of the food colours could be determined. Sunset Yellow, which is a 1-phenylazo-2-naphthol dye, gave an E_p of − 0.66 V at an optimum pH of 10 and a corresponding current of 400 nA for a 10^{-6} M concentration after 10 min accumulation at the HMDE. Tartrazine [sodium 5-hydroxy-1-*p*-sulphophenyl-4-(*p*-sulphophenylazo)pyrazole-3-carboxylate] gave rise to a similar phenomenon at the HMDE and with the addition of 100 ppm of tetraphenylphosphonium chloride (TPPC) shifted the reduction potential to a more negative value while increasing the height of the peak. This phenomenon was used for partial identification of tartrazine in a pharmaceutical tablet. The chosen tablet was stirred with 10 ml of water until the colour had dissolved. The solution was separated and diluted to 20 ml with water. A 0.5 ml aliquot of this solution was added to 19.5 ml of pH 10 buffer (10^{-2} M in carbonate, 5×10^{-2} M in KCl) and the solution was analysed by AdSV. Two peaks were obtained, and the second one, at − 0.74 V (vs Ag/AgCl), on addition of 100 ppm of TPPC, was shifted to − 0.84 V with doubling of its peak height. This amounts to partial confirmation of tartrazine. Becerro Dominguez *et al.* (1990) recently used DPP to determine these azo dyes, Sunset Yellow and tartrazine, in soft drinks. The azo dye 3-(4,5-dimethyl-2-thiazolylazo)-2,6-diaminopyridine has been determined by Alonso *et al.* (1987) by DPAdSV at an HMDE with an LOD of 8×10^{-10} M for 4 min accumulation in phosphate buffer (pH 7.0).

Miranda Ordieres *et al.* (1987) investigated the electrochemical behaviour of the N=N-containing anticancer drug dacarbazine [5-(3,3-dimethyl-1-triazenyl)imidazole-4-carboxamide] (**IX**) by Tast polarography and DPP at the DME, by cyclic and differential-pulse voltammetry at the HMDE and by anodic voltammetry at the glassy carbon electrode (GCE). Calibration graphs were obtained in the ranges 2×10^{-8}–2×10^{-5} M by DPP, 5×10^{-9}–1×10^{-5} M by AdSV and 1×10^{-5}–1×10^{-4} M by HPLC–oxidative ED. Serum spiked with ng ml^{-1} concentrations of the drug down to a limit of detection of 18 ng ml^{-1} could be determined by HPLC–ED (Fig. 17).

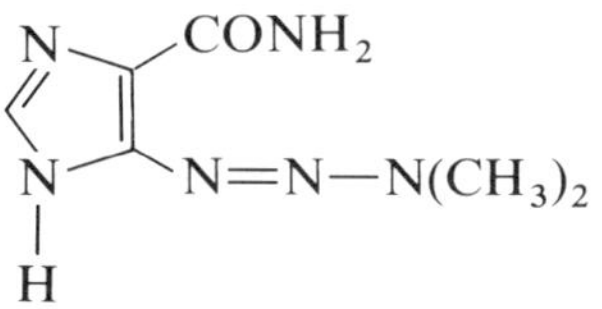

(**IX**)

Azo compounds can chelate with a wide range of metal ions, as illustrated in Table 12. The resulting chelates can be adsorbed on stationary electrode surfaces such as the HMDE and the metals determined by AdSV. This has found application in analyses of sea water (Wang *et al.*, 1987a), snow (Wang *et al.*, 1985b), rice (Lu *et al.*, 1990a), ore samples (Lu *et al.*, 1990b) and human hair (Zhao and Jin, 1988, 1989). Table 13 shows the application of the chelating agent 5-Br-PADAP to the determination of iron in tap water, HCl and human hair (Zhao and Jin, 1989).

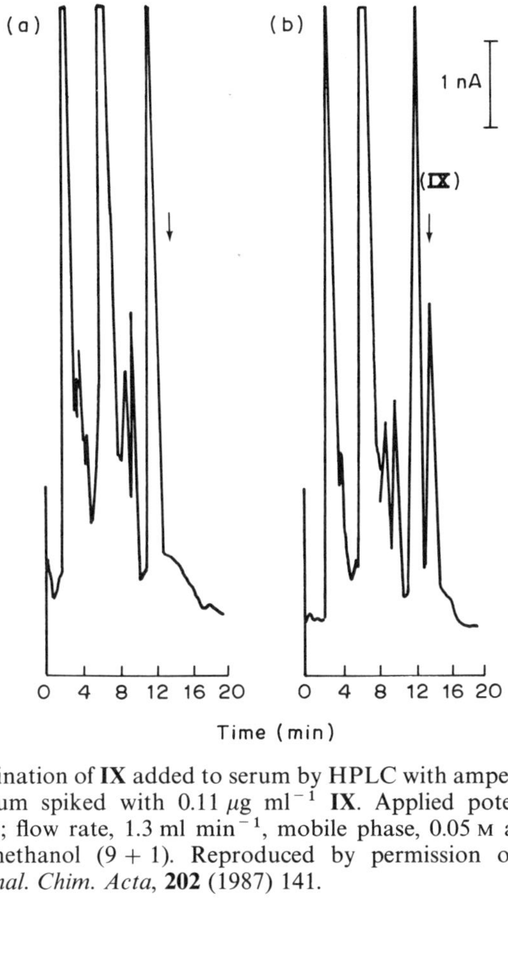

Figure 17. Determination of **IX** added to serum by HPLC with amperometric detection: (a) blank; (b) serum spiked with 0.11 μg ml^{-1} **IX**. Applied potential, + 0.98 V (vs Ag/AgCl/sat. KCl); flow rate, 1.3 ml min^{-1}, mobile phase, 0.05 M ammonium acetate buffer (pH 5.5)–methanol (9 + 1). Reproduced by permission of Elsevier Science Publishers from *Anal. Chim. Acta*, **202** (1987) 141.

1.6 NITROGEN-CONTAINING AROMATIC HETEROCYCLES

1.6.1 HETEROCYCLES CONTAINING ONE NITROGEN ATOM

The polarographic behaviour of heterocyclic nitrogen compounds has been reviewed by Zuman and Perrin (1969). Compounds containing only onc nitrogen atom in an aromatic ring system such as pyridine and quinoline generally only produce catalytic hydrogen waves in aqueous solutions of pH < pK_a. Substituted pyridine compounds, such as those containing a COOH group in the 2-position, give rise to well defined waves or peaks for analytical purposes, e.g. the pesticide picloram (potassium 4-amino-3,5,6-trichloropicolinate) has been determined down to a limit of detection of 0.02 ppm (Gilbert and

Table 12. AdSV data for the determination of metal ions after chelation with an azo compound

Metal	Azo compound	Linear calibration range (M)	Limit of detection (M)	Ref.
Zr	Solochrome violet RS	1.1×10^{-8}–1.1×10^{-7}	2.3×10^{-10}[a]	Wang *et al.* (1987a)
Al	Solochrome violet RS	1–30 $\mu g\ l^{-1}$	0.15 $\mu g\ l^{-1}$	Wang *et al.* (1985b)
Fe(III)	Solochrome violet RS	—	1 $\mu g\ l^{-1}$	Hua (1988)
Cr(III)	5-Br-PADAP[b]	3×10^{-9}–9×10^{-8}	9×10^{-10}	Lu *et al.* (1990a)
Cu(II)	5-Br-PADAP	5×10^{-8}–5×10^{-7}	—	Tanaka *et al.* (1990)
V(V)	5-Br-PADAP	—	5×10^{-10}	Lu *et al.* (1990b)
Bi	5-Br-PADAP	1×10^{-10}–6×10^{-8}	7.5×10^{-11}	Zhao and Jin (1988)
Fe(III)	5-Br-PADAP	2.5×10^{-10}–1×10^{-7}	—	Zhao and Jin (1989)
Ti(IV)	5-Br-PADAP	4×10^{-10}–5×10^{-8}	3×10^{-10}	Jinkui and Neeb (1990a)
Ti(IV)	PAR[c]	0.15–4.8 ng g^{-1}	0.1 ng g^{-1}	Jinkui and Neeb (1990b)
Bi(III)	PAN[d]	—	—	Dong *et al.* (1991)

[a]For 10 min preconcentration.
[b]5-Br-PADAP = 2-(5′-bromo-2′-pyridylazo)-5-diethylaminophenol.
[c]PAR = 4-(2-pyridylazo) resorcinol.
[d]PAN = 1-(2-pyridylazo)-2-naphthol.

Mann, 1973) and Filimonova and Gorbunova (1973) determined the same pesticide in soil and dried maize down to concentrations of 0.7 and 0.25 $\mu g\ ml^{-1}$, respectively. The electrochemical reaction involved loss of the protonated 4-amino group from the substituted pyridine molecule to give NH_3.

Pyridines substituted by COOH in the 3- or 4-position undergo a different

Table 13. Results for the determination of iron in samples of HCl, tap water, and human hair

Sample	Fe content determined by AdSV (ppm)	Average (ppm)	Fe content determined by AAS (ppm)
HCl	0.47, 0.46, 0.47, 0.44, 0.45	0.46	0.48
Tap water	2.3, 2.6, 2.5, 2.4, 2.4	2.4	2.6
Hair A	37.8, 38.3, 37.4, 39.6, 37.2	38.1	38.9
Hair B	44.4, 44.6, 45.8, 46.3, 43.5	44.7	44.9
Hair C	113, 117, 114, 118, 112, 115	115	119

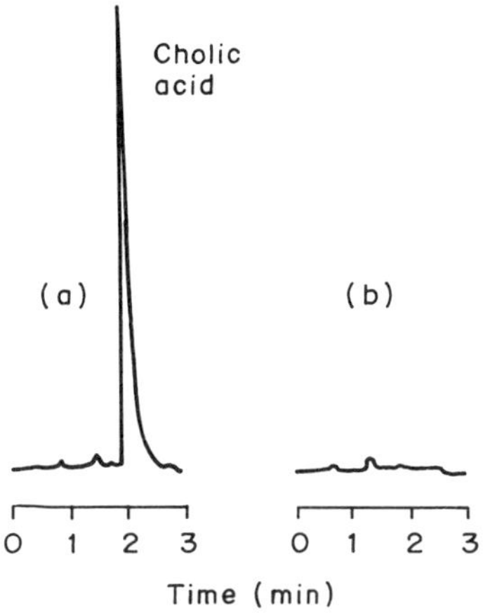

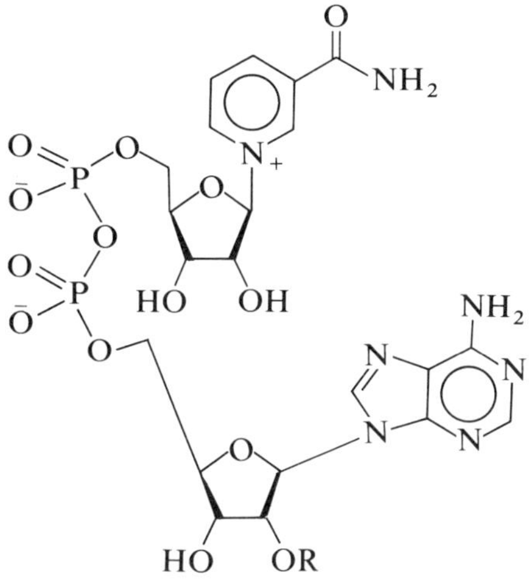

(X)

Figure 18. Chromatograms of cholic acid (100 pmol on-column) detected via enzymatic NADH production at + 0.8 V vs Ag/AgCl at (a) base-activated and (b) untreated glassy carbon electrodes. Mobile phase, 0.1 mM NAD in methanol–0.05 M phosphate buffer (pH 7.5) (30 + 70); flow rate, 1.5 ml min^{-1}, temperature, 30°C. Reprinted with permission from *Analytical Chemistry*, **63**, 845. Copyright (1991) American Chemical Society

mechanism of reduction from the 2-substituted species, involving saturation of an endocyclic C=N group. Vallon *et al.* (1975) determined isonazid and isonicotinic acid using a.c. polarography down to 0.2 and 0.03 μg ml^{-1}, respectively, and de Silva *et al.* (1973) determined 2-hydroxynicotinic acid and its major metabolite in blood and urine by DPP. Other heterocyclic compounds containing one nitrogen atom in their ring system, e.g. piperidines and indoles, are not reduced at the DME although they can exhibit oxidation phenomena at solid electrodes, e.g. involving *N*-oxidation.

Eisenberg and Cundy (1991) showed the utility of an electrochemically base-activated glassy carbon electrode as a sensor for NADH in HPLC–ED. The treated electrode was compatible with mobile phases commonly used in HPLC. By being able to detect NADH, it was possible to follow chemical entities involved in the NAD^+ (**X**)–NADH-coupled oxidation processes. Figure 18 shows chromatograms of cholic acid detected via enzymatic NADH production at + 0.8 V (vs Ag/AgCl) at both base-activated and untreated glassy carbon electrodes.

1.6.2 HETEROCYCLES CONTAINING TWO NITROGEN ATOMS

Heterocyclic aromatic compounds containing two nitrogen atoms in the ring system can be reduced at the DME, e.g. pyrimidines and pyrazines. The reduction process of pyrimidines involves saturation of an endocyclic C=N group followed by formation of tetrahydropyrimidine. The pyrimidine-containing drug trimethoprim, used with a sulphonamide as a combination antibacterial drug, can be determined by reductive polarography in the presence of the non-reducible sulphonamide (Smyth, 1989). Further, DPP of this combination drug formulation, in pH 4.1 buffer, showed additional peaks at − 0.15, −0.27 and − 0.87 V in addition to the trimethoprin peak at − 1.32 V, indicative of process impurities and/or degradation products.

Brooks *et al.* (1973a) used DPP to determine trimethoprim and its *N*-oxide metabolites in urine. It is instructive to compare this analytical method with rivals for the determination of trimethoprim in a biological matrix such as urine (Table 14).

Gao *et al.* (1990) used chemically modified carbon paste electrodes containing 1,10-phenanthroline and Nafion as modifiers in order to preconcentrate Fe(II) from dilute solutions prior to DPV. The method has been used for the determination of iron in certified standard reference materials and at trace levels in natural waters.

1.6.3 HETEROCYCLES CONTAINING THREE NITROGEN ATOMS

Heterocyclic compounds containing three nitrogen atoms are also reducible at the DME owing to reduction of endocyclic C=N groups. Thompson (1965) investigated the polarographic behaviour of several 1,3,5-triazine herbicides and found that they mostly could be reduced in acidic supporting electrolytes with detection limits of 2×10^{-7} M by single-sweep polarography. Hernandez Mendez *et al.* (1985) investigated the behaviour of the pesticide menazon (**XI**) and its hydrolysis products. As can be seen from its structure (Fig. 19), (**XI**) is a combination of a thiophosphate and a triazine and will hydrolyse by fission of the P—S bond to yield 4,6-diamino-1,3,5-triazin-2-methanethiol (**XII**) and the corresponding thiophosphate. Compound **XI** gives two diffusion-controlled cathodic peaks at pH 4.78 due to fission of the C—S bond followed by saturation of a C=N group. This latter peak at − 0.80 V (vs Ag/AgCl) can be

Table 14. Comparison of analytical methods for the determination of trimethoprim in urine

Analytical technique	Time per assay	Selectivity of method	Limit of detection	Ref.
HPLC-UV	Relatively rapid	Good	Good	Vree (1978)
DPP	Moderate	Moderate	Good	Brooks *et al.* (1973a)
GC-FID	Slow	Good	Moderate	Kuftinec (1978)
Visible spectrophotometry	Slow	Poor	Poor	Rao (1978)

used for DPP determination of **XI** down to 5.5 × 10^{-7} M and the hydrolysis products (**XII–XIV**) can also be detected by DPP down to 1.7 × 10^{-7} M. This scheme of hydrolysis and reduction is illustrated in Fig. 19.

McKone *et al.* (1972) compared GC, polarographic and spectrophotometric methods for the determination of terbutryne, ametryne and atrazine in pond and canal water. The polarographic method involved extraction with dichloromethane, taking up the residue in 0.005 M H_2SO_4–50% methanol and running the linear sweep voltammogram from − 0.8 V to the decay of the supporting electrolyte. In this medium, atrazine gave a peak at − 1.05 V with a limit of detection of 0.01 ppm, whereas terbutryne and ametryne were both reduced at − 1.45 V with a limit of detection of 0.005 ppm. The polarographic method matched the GC procedure in terms of time of analysis and ease of calculation of results, but was inferior in terms of limit of detection and selectivity. Capillary column GC–MS (Dietrich *et al.*, 1988) has been applied to the identification and determination of synthetic organic chemicals in a river in North Carolina, USA.

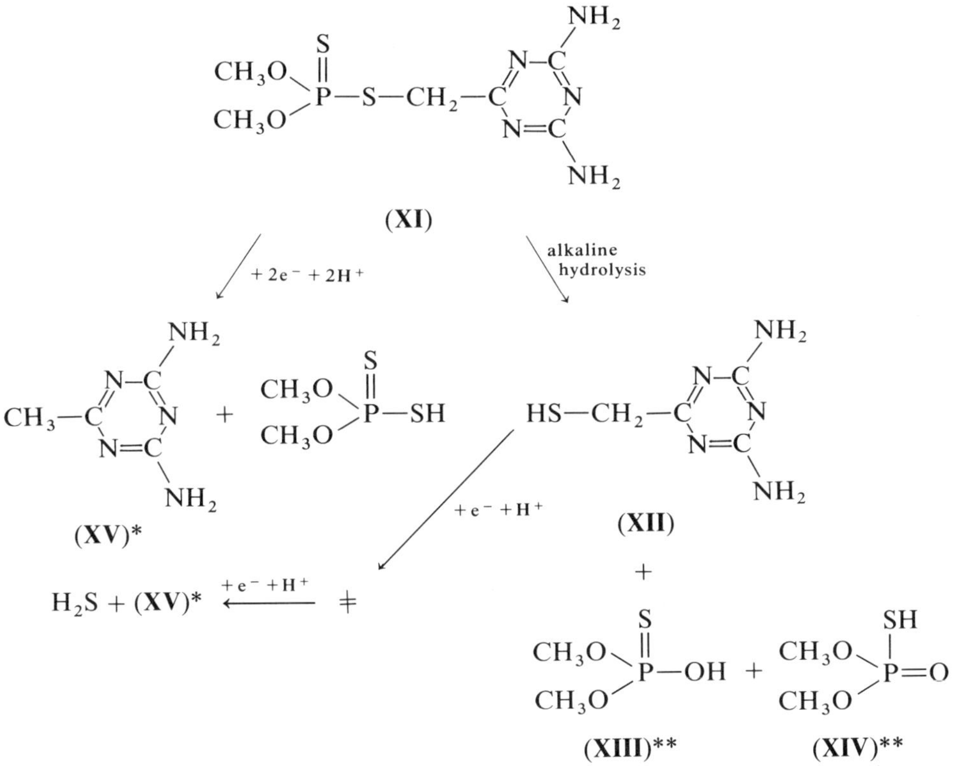

Figure 19. Scheme of hydrolysis and polarographic reduction for the pesticide menazon, *S*-[(4,6-diamino-1,3,5-triazin-2-yl)methyl]-*O,O'*-dimethylphosphorodithioate. *Can be reduced further by a 2e⁻ process (saturation of the C=N group). **Give rise to anodic phenomena at the DME

Sample preparation was by either resin adsorption or dichloromethane liquid–liquid extraction. Of the 48 distinct synthetic organic chemicals detected, those found most frequently included the triazine herbicides atrazine and methylatraton. Atrazine was positively identified in samples at low ppb concentrations, showing the method's superiority over polarographic methods for these molecules.

Goicolea *et al.* (1991) have recently studied the AdSV behaviour of triazine herbicides metamitron (**XVI**) and isometiozine and found LODs of 3×10^{-10} and 10^{-9} M, respectively, at reduction potentials -0.45 and -0.56 V, respectively. They could be determined in soils by this technique following extraction with dichloromethane and at levels of 0.44 and 0.97 ng g^{-1} for metamitron and isometiozine, respectively.

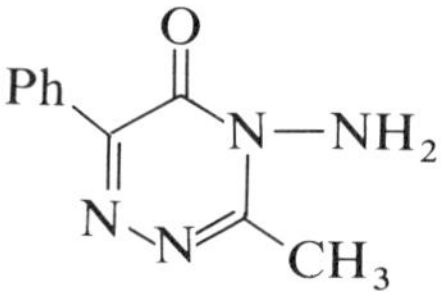

(**XVI**)

1.6.4 HETEROCYCLES CONTAINING FOUR NITROGEN ATOMS

Riboflavin (vitamin B_2) contains four nitrogen atoms in a tricyclic ring system and can undergo controlled adsorptive accumulation at the SMDE (Wang *et al.*, 1985d), prior to DPV, with a limit of detection of 2.5×10^{-11} M after 30 min preconcentration. This limit of detection is about three orders of magnitude lower than that obtained by solution-phase DPP and is also hard to match by rival analytical techniques. However, in terms of the selectivity required for the determination of this vitamin in multivitamin preparations, then a solely electroanalytical method such as AdSV is found somewhat lacking. For example, Amin and Reusch (1987) devised an ion-pair reversed-phase HPLC method for the separation and simultaneous determination of riboflavin, thiamine hydrochloride (vitamin B_1), nicotinamide, pyridoxine hydrochloride (vitamin B_6) cyanocobalamin (vitamin B_{12}), ascorbic acid (vitamin C) and folic acid in capsules. These seven substances were extracted from the preparation in 5 min, followed by HPLC on a LiChrosorb RP-18 column using methanol–water–phosphoric acid–octanesulphonic acid as the eluent with UV detection at 254 nm. All of the substances were separated in less than 4 min with capacity factors (k') ranging from 6.0 to 19.7. In a biological fluid such as haemodialysate, riboflavin can be satisfactorily determined by HPLC with fluorescence detection in the concentration range 2–20 ng ml^{-1} (Mohammed *et al.*, 1981).

Folic acid, which also contains four nitrogen atoms in a tricyclic ring system, can be determined, along with five other water-soluble vitamins, by HPLC–ED, using a parallel dual-electrode electrochemical detector (Hou and Wang, 1990a). One electrode was held at + 0.80 V and the other at + 1.20 V (vs SCE) and, under these conditions, eight other water-soluble vitamins, including riboflavin, did not interfere. The linear calibration range was found to be 5–200 ng for folic acid with a limit of detection of 2 ng.

Folic acid can be determined with a lower LOD but decreased selectivity by solely AdSV (Luo, 1986). If a series dual electrochemical detector is used with the upstream electrode held at − 0.40 V (vs SCE) and the downstream electrode at + 0.10 V (vs SCE), then riboflavin can be determined in a linear calibration range of 4 ng–2.6 μg with a limit of detection of 4 ng. This method (Hou and Wang, 1990b) was found to be selective for riboflavin in the presence of 13 other vitamins. Hou and Wang (1990c) also applied this technique to the selective determination of flavin adenine dinucleotide (FAD).

Eisenberg *et al.* (1990) used a novel HPLC method with an immobilised xanthine oxidase reactor and an electrochemical detector for the simultaneous determination of allopurinol (4-hydroxypyrazolo[3,4-*d*]pyrimidine) and its major metabolite, 3,4-dihydroxypyrazolo[3,4-*d*]pyrimidine, in rat plasma, intestinal wash and bile. Xanthine oxidase was immobilised on 5 μm aldehyde silica (prepacked into a 10 mm × 2 mm i.d. cartridge) in a simple procedure. Allopurinol eluted from an analytical column was converted to oxypurinol in the enzyme reactor with the eluent as the reaction medium and detected with a high degree of selectivity with an electrochemical detector at + 0.85 V.

High specificity of the enzymatic reaction combined with the selectivity of the ED eliminated the need for extensive sample preparation. The assay was linear in the range 15–500 ng ml^{-1} for rat plasma, intestinal wash and bile with a limit of detection of 10 pg on-column (signal-to-noise ratio = 4) for both allopurinol and oxypurinol.

1.7 AMINES

1.7.1 PRIMARY AMINES

Primary aliphatic amines such as hydrazine and 1,1-dimethylhydrazine can be monitored in industrial hygiene situations by HPLC–ED after derivatisation with salicylaldehyde, as shown for hydrazine (**XVII**) (Kester and Danielson, 1984). Less than 5 ng of the resulting derivatives could be detected using electrooxidation of the phenolic group. The limit of detection for (**XVII**) was estimated to be 0.025 ppm.

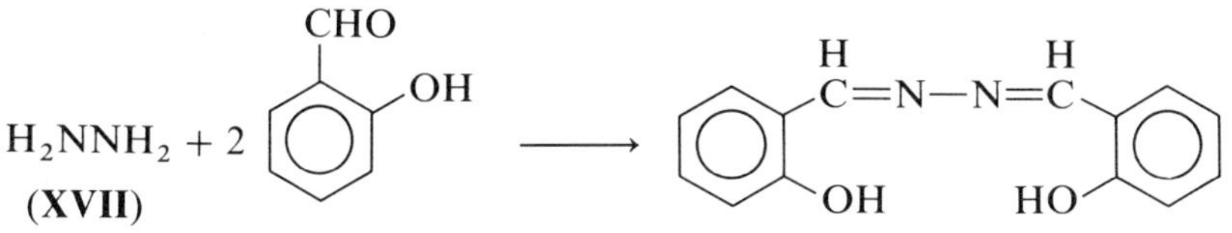

Primary aromatic amines such as 4,4′-methylenebis(2-chloroaniline) (MOCA) and 2-chloroaniline (OCA) can be determined in factory atmospheres by HPLC–ED without resort to derivatisation. Purnell and Warwick (1980) employed Tenax GC as the adsorbent in the sampling device to collect amine vapours and a filter was used to collect particulate amines. The ED was found more selective and had a lower LOD than the UV detector and MOCA and OCA could be determined down to 20 and 10 pg injected, respectively. The EPA priority pollutants benzidine and 3,3′-dichlorobenzidine are conveniently determined at low ppb concentrations by reversed-phase HPLC–ED using a tubular carbon black–polyethylene working electrode held at + 0.70 V (Fig. 20). Again the LOD was significantly lower than with UV detection (Armentrout and Cutie, 1980).

The potent mutagens 1- and 2-aminoanthracene present in petroleum substitutes can also be determined by HPLC–ED. Judicious choice of the potential of the detector can selectively determine them in the presence of other polycyclic aromatic hydrocarbons and polycyclic aromatic nitrogen heterocycles which are oxidised at more positive potentials (Caudill *et al.*, 1983).

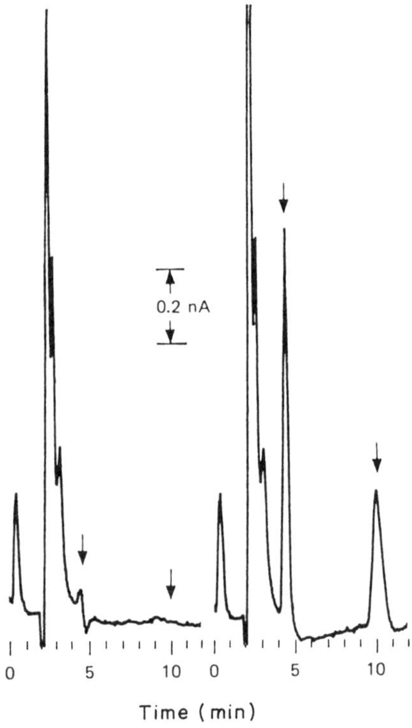

Figure 20. LC–ED of wastewater before and after spiking with 2.7 ppb benzidine and 2.9 ppb dichlorobenzidine. Reproduced from the *Journal of Chromatographic Science* by permission of Preston Publications, A Division of Preston Industries, Inc.

Selenium(IV) can be determined by CSV after the formation of a piazselenol with 3,3′-diaminobenzidine (Breyer and Gilbert, 1987). The Se was then accumulated as HgSe at an HMDE held at – 0.45 V. Differential-pulse CSV gave a detection limit of 0.01 ppb. For the determination of Se(IV) in natural waters, interferences can be avoided by extraction of the piazselenol into toluene followed by back-extraction into 0.5 M HCl. The accuracy of the method was checked by analyses of a standard reference material and applied to the determination of Se(IV) in sea-water samples at concentrations of the order of 20 ng l^{-1} with a concentration factor of ten during the extraction procedure. This electroanalytical method compares very favourably with hydride generation–ICP–MS, which has LODs of 34 pg ml^{-1} for ^{77}Se and 26 pg ml^{-1} for ^{82}Se for the determination of Se in sediment reference material (McCurdy *et al.*, 1991). Stara and Kopanica (1988) also used 3,3′-diaminobenzidine to give a working method for Se(IV) with a linear calibration graph in the range 3×10^{-9}–3×10^{-8} M. Wang *et al.* (1990) found a limit of detection of 4×10^{-10} M for Se(IV) using AdSV of its complex with phenylenediamine.

1.7.2. SECONDARY AMINES

Secondary amines are susceptible to nitrosation procedures, although the electrochemical methods based on these procedures have generally found application only in formulation analysis. An exception is the determination of the herbicide glyphosate (**XVIII**) in natural waters (Bronstad and Friestad, 1976). The analytical method involved clean-up of the water sample and the preconcentration of the analyte on an ion-exchange column followed by its nitrosation directly in the eluate. The LOD for DPP determination of glyphosate was quoted as 35 ng ml^{-1}. Sample polarograms are given in Fig. 21.

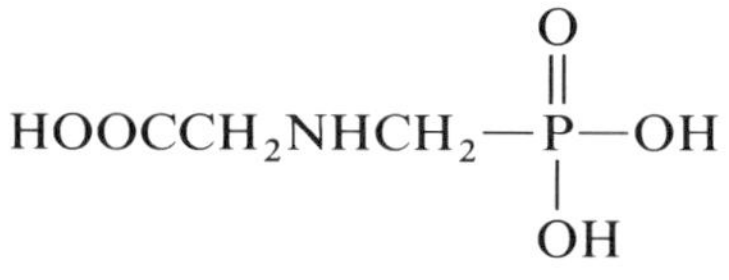

(**XVIII**)

1.7.3 TERTIARY AMINES

Tertiary amines are susceptiblc to *N*-oxidation procedures and these derivatisations again have mostly been applied in formulation analysis. Alternatively, Rodriquez *et al.* (1990) produced a direct voltammetric assay for the tertiary amine heroin in illicit dosage forms, utilising its oxidation at a carbon paste electrode. The results were in good agreement with an alternative HPLC assay. Schwartz and David (1985) used HPLC–ED with the detector set at + 1.20 V (vs Ag/AgCl) to determine heroin, morphine and cocaine with LODs of 1, 0.3

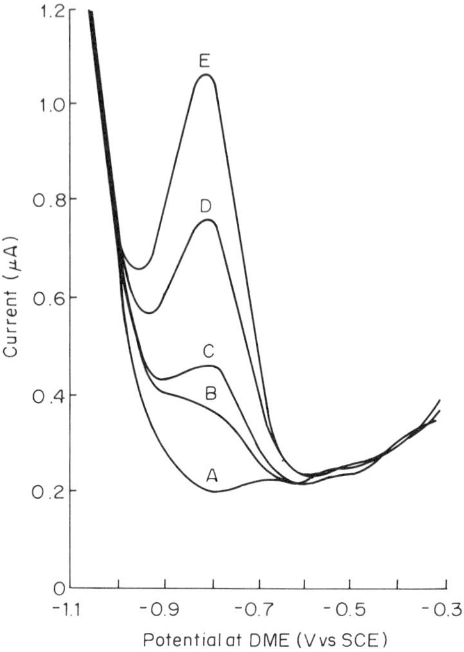

Figure 21. Current–potential graphs for glyphosate nitrosamine after anion-exchange treatment and nitrosation of various amounts of glyphosate added to 1 l of local tap water. A, 0 (reagent blank); B, 35; C, 70; D, 140; and E, 210 μg added. Reproduced by permission of the Royal Society of Chemistry

and 2 ng, respectively. Moncrieff (1990) also used HPLC–ED for the simultaneous determination of the natural alkaloid sparteine and its 2-dehydro and 5-dehydro metabolites. The tertiary amine methadone (**XIX**) has been determined in urine by linear-sweep voltammetry and differential-pulse voltammetry by Rodriquez *et al.* (1989) with an LOD of 0.3 μg ml^{-1}. The determinations were in good agreement with those of an EMIT assay.

Wang and Freiha (1983) investigated selective voltammetric detection based on adsorptive preconcentration for flow-injection analysis (FIA). They found the technique to be highly selective compared with conventional amperometric detection, owing to the combination of intimate contact between the surface-bound species and the electrode surface with the medium exchange procedure.

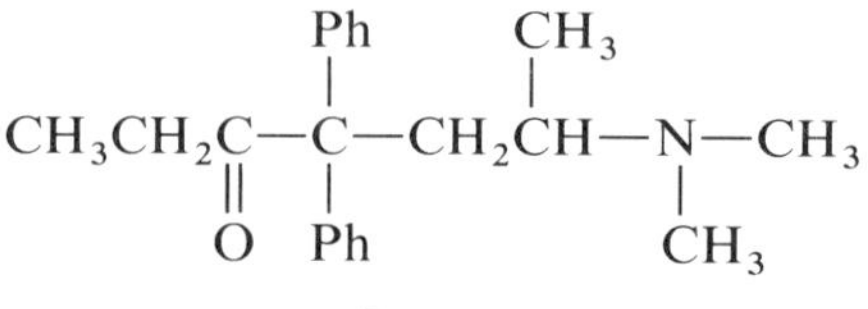

(XIX)

This was demonstrated by the determination of the tertiary amine chlorpromazine in 100-fold excesses of non-adsorbable solution species with similar redox potentials. With differential-pulse quantification, LODs of a few nanograms were obtainable and injection rates of 24 h^{-1} were demonstrated. Reproducible determination of chlorpromazine in urine was found possible with no sample pretreatment. Belal and Anderson (1985) were able to achieve a sampling rate as high as 200 h^{-1} using a flow rate of 1 ml min^{-1} for the determination of perphenazine, triflupromazine and fluphenazine by FIA–ED. Table 15 gives the results of the assay of dosage forms by this method. These results compared favourably with those given by the USP method of non-aqueous titration and application of the variance ratio test and Student's *t*-test showed no significant difference between the two methods.

Duthu (1984) developed a selective method with a low limit of detection for the determination of serum concentrations of erythromycin A by HPLC–ED. Erythromycin was extracted from alkalinized serum samples with methyl *tert*-butyl ether. After evaporation of the ether, the samples were reconstituted in acetonitrile–sodium perchlorate–ammonium acetate–methanol under isocratic conditions. Eluted peaks were detected by dual coulometric electrodes operated in the oxidative screen mode. In this manner, many components in the extracted serum samples were irreversibly oxidised at the upstream electrode held at + 0.7 V and did not contribute to the detected current at the downstream electrode at + 0.8 V. The recovery of erythromycin from serum was 84 %, the limit of determination 0.05 μg ml^{-1} in serum and the dynamic linear range 0.05–1.5 μg ml^{-1}. This method was used to determine both erythromycin and its gastric degradation products in human serum and, in addition, other macrolide antibiotics could be measured with ED (Table 16). Only the two *N*-desmethyl erythromycins were found to be undetectable. Cyclic voltammetry on the glassy carbon electrode confirmed the absence of an oxidation peak below + 1.3 V for these two compounds, suggesting the importance of the tertiary amine of the desosaminyl sugar (**XX**) in determining an oxidation signal for the molecule.

Table 15. Assay of dosage forms of phenothiazines by FIA–ED

	Recovery (%)[a]	
Preparation	FIA–ED method	USP method
Fluphenazine. 2HCl tablets (4 mg per tablet)	98.50 (0.45)	99.1 (0.72)
Perphenazine tablets (5 mg per tablet)	99.2 (0.37)	99.5 (0.83)
Triflupromazine. HCl tablets (8 mg per tablet)	100.7 (0.54)	100.9 (0.91)

[a]The results given are the averages of six separate determinations, with relative standard deviations (%) in parentheses.

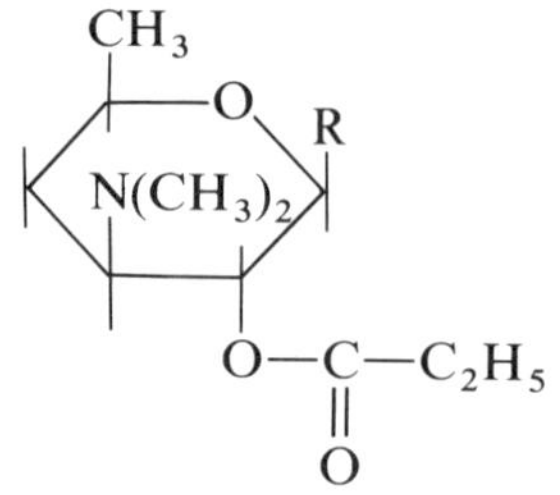

(XX)

1.7.4 QUATERNARY AMINES

Quaternary ammonium moieties that are attached to an aromatic ring can be reduced at the DME as follows:

$$Ar\overset{+}{N}H_3 + 2e^- + H^+ \rightarrow ArH + NH_3$$

This mechanism has been shown to be responsible for the first portion of the first wave exhibited by tetracycline hydrochloride in acidic solution (Caplis, 1970) and has also been postulated as contributing to the electrochemical behaviour of the pesticide, picloram (Filimonova and Gorbunova, 1973). The pesticide diquat gives well defined waves in 0.1 M HCl (Nangniot, 1970) at − 0.56 and − 1.06 V (vs SCE) due to two $1e^-$ reductions. Diquat and the related molecule paraquat, both of which contain two quaternary nitrogens, can be differentiated in mixtures based on their reduction potentials in 0.08 M HCl. Nangniot (1970) further described a method for the determination of these

Table 16. Relative retention times and relative peak areas of various macrolide antibiotics and erythromycin derivatives

Macrolide antibiotics	Relative retention time[a]	Relative peak area	Erythromycin A derivatives	Relative retention time[a]	Relative peak area
			Descladinose	0.59	1.70
Josamycin	0.66	0.94	*N*-Demethyl-	0.88	ND[b]
Oleandomycin	0.88	0.79	*N*-Didemethyl-	0.80	ND[b]
			2-Propionate	1.01	0.75
Erythromycin A	1.00	1.00	2-Ethylsuccinate	1.00	0.92
Erythromycin B	1.19	0.93			
Erythromycin D	1.02	1.04	Dihydro-	0.96	1.02
Tylosin	0.51	0.69	Anhydro-	1.26	1.06
			Enol ether	2.70	1.16
			Spiroketal	1.25	1.66

[a]Retention times were determined by UV detection at 214 nm for a 10 μg injection.
[b]N.D. = not detected by electrochemical detector.

compounds in residues following extraction with sulphuric acid–octan-2-ol, clean-up and separation on Celite 545 and Zeo-Karb 225 columns followed by polarographic determination in the above-mentioned medium.

Webb and Johnson (1986) developed a simple, specific and sensitive method for the determination of plasma choline (**XXI**). The eluent from a reversed-phase column is mixed with choline oxidase in a post-column reaction coil to produce H_2O_2, which is monitored electrochemically by oxidation at a Pt electrode:

$$(CH_3)_3\overset{+}{N}CH_2CH_2OH + 2\ O_2 + H_2O \xrightarrow[\text{oxidase}]{\text{choline}} \text{betaine} + 2\ H_2O_2$$

(**XXI**)

Choline can be measured in plasma samples containing less than 1 μmol l^{-1} of the compound and the method was found to give a linear calibration graph in the range 1–20 μmol l^{-1}. Catecholamines and ascorbic acid were found not to interfere.

1.7.5 POLYAMINES

Guette and Rips (1988) reported a rapid and simple method for the determination of polyamines in cerebral fractions, without prior purification, by reversed-phase HPLC–ED. Precolumn derivatisation with *o*-phthalaldehyde–ethanethiol at pH 9.35 for 5 min was employed prior to HPLC on Nucleosyl CN (5 μm) and ED at + 0.70 V. The internal standard was 3,3′-diaminopropylamine and the polyamines spermine, putrescine and spermidine were determined (Fig. 22).

1.8 AMINO ACIDS AND PEPTIDES

Recent progress in amino acid determination can be attributed to technological advances in HPLC and chromatographic detectors. Both cation- and anion-exchange stationary phases are used together with a variety of derivatisation procedures to overcome the lack of response of traditional HPLC detectors towards amino acids. For example, Oates and Jorgenson (1989) used open-tubular HPLC–ED to separate and determine 18 naphthalene-2,3-dicarboxaldehyde-labelled amino acids. The usefulness of the technique was demonstrated by the determination of so-labelled hydrolysis products from bovine chymotrypsinogen.

Welch *et al.* (1990) used pulsed amperometric detection to avoid the derivatisation step, and this resulted in a relatively low level of detection for amino acids but a limited linear range of response. Such pulsed amperometric detection uses a triple-step potential wave form to combine amperometric detection with alternating anodic and cathodic polarisations to clean and reactivate the electrode surface. This waveform exploits the surface-catalysed oxidation of the amine group, activated by the transient formation of surface oxides on noble

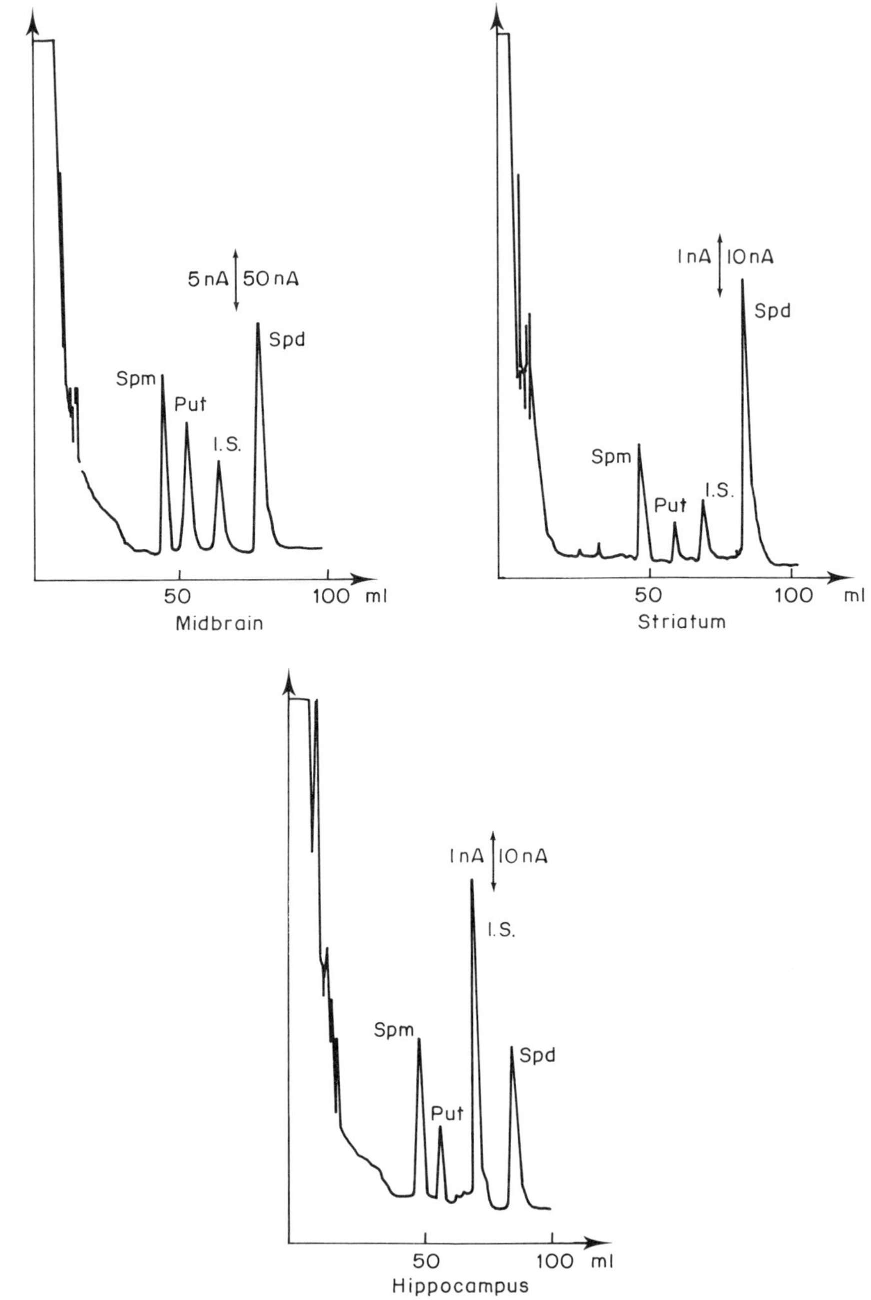

Figure 22. Determination of polyamines in biological samples. Mouse brain fragments: cerebellum, midbrain, pons, medulla oblongata, frontal cortex, hypothalamus, thalamus, hippocampus, striatum, remains. Products measured = spermine (Spm), putrescine (Put), internal standard (I.S.), spermidine (Spd). Reproduced by permission of Plenum Publishing Corporation; from Rips and Guette, in *Contemporary Electroanalytical Chemistry* (ed. Ivaska, A.), p. 301

metals. Zoulis *et al.* (1990) indirectly determined L-tryptophan in serum by cleavage to indole with tryptophanase followed by AdSV of the indole at a carbon paste electrode. The medium exchange procedure is used and concentrations of L-tryptophan in the range 20–80 μM can be determined. This method avoids the formation of coloured compounds between indole and *p*-dimethylaminobenzaldehyde (Erlich reaction), as required by spectrophotometric methods. The method was also found to be faster and cheaper than HPLC procedures, where a deproteinisation step is usually necessary. Melatonin is synthesised from typtophan biochemically and Azekawa *et al.* (1990) were able to measure this former molecule in microdialysates from the pineal gland of the rat by HPLC–ED with a limit of detection of 5 pg.

Peptides longer than dipeptides react in the classical biuret reaction with Cu(II) to yield electroactive Cu(II)–peptide complexes that can be oxidised to the corresponding Cu(III) complexes. This allows for the development of reversed-phase HPLC–ED methods to determine low concentrations of peptides that lack an electroactive group or a primary amine. Warner and Weber (1989) found the selectivity for a model peptide to be 10^3–10^4 over non-electroactive amino acids.

Shibanoki *et al.* (1990) used HPLC–ED with the detector set at + 1.05 V (vs Ag/AgCl) to determine [Leu]- and [Met]-enkephalin and their *N*-terminal tyrosine-containing metabolic fragments (Tyr, Tyr–Gly, Tyr–Gly–Gly and Tyr–Gly–Gly–Phe). This procedure was applied to an investigation of the activities of enkephalin-degrading enzymes in whole plasma collected from rats, mice and chicks. Table 17 compares the formation of Tyr, Tyr–Gly and Tyr–Gly–Gly from [Leu]-enkephalin for the three species in the presence and absence of bestatin, an inhibitor with high selectivity for aminopeptidase M.

Table 17. Enzymatic formation of Tyr, Tyr–Gly and Tyr–Gly–Gly from [Leu]-enkephalin in mouse, chick and rat plasma. Hydrolysis was measured 5 min after substrate addition (64.8 μM) in the presence and absence of bestatin (500 μM)

		Product (nmol ml^{-1} plasma)[a]		
Species	Inhibitor	Tyr	Tyr–Gly	Tyr–Gly–Gly
Mouse				
	None	19.7 ± 4.8	0.33 ± 0.05	2.90 ± 2.16
	Bestatin	1.3 ± 1.4	0.73 ± 0.14	8.41 ± 4.50
Chick				
	None	68.7 ± 2.9	ND	0.11 ± 0.05
	Bestatin	5.3 ± 4.7	ND	1.82 ± 0.24
Rat				
	None	59.7 ± 6.0	0.46 ± 0.09	0.83 ± 0.14
	Bestatin	3.4 ± 3.1	1.56 ± 0.25	3.24 ± 0.18

[a]Values are means ± S.D. ($n = 5$); ND = not detectable.

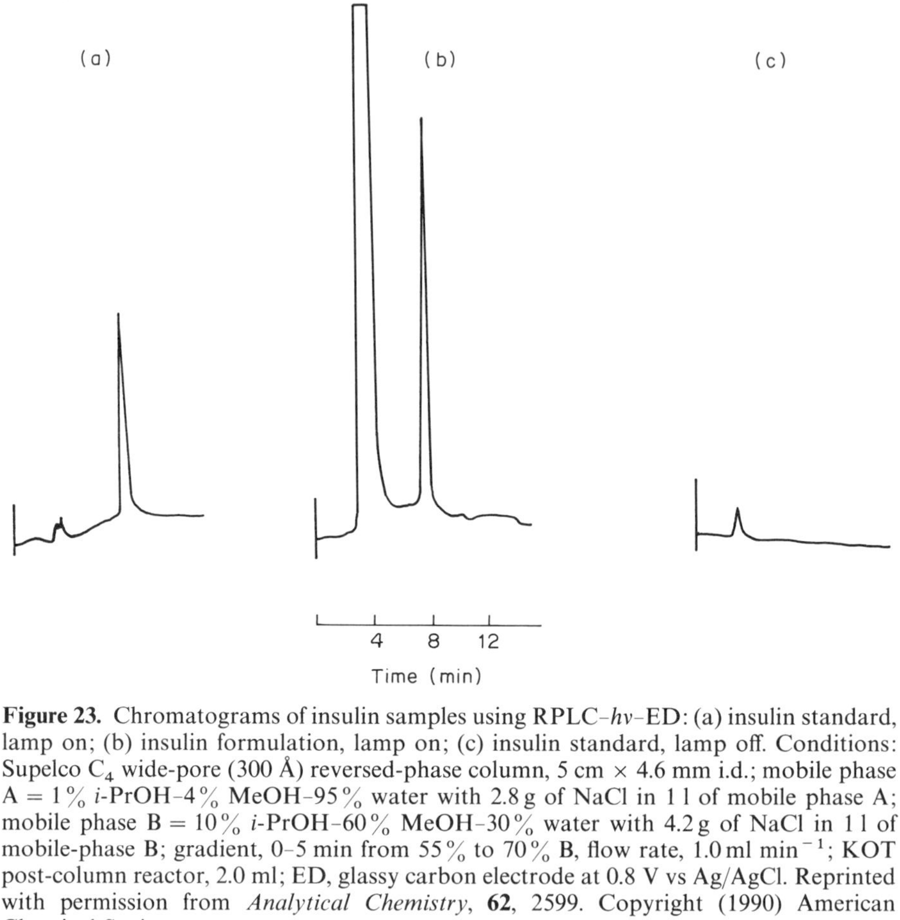

Figure 23. Chromatograms of insulin samples using RPLC–*hv*–ED: (a) insulin standard, lamp on; (b) insulin formulation, lamp on; (c) insulin standard, lamp off. Conditions: Supelco C_4 wide-pore (300 Å) reversed-phase column, 5 cm × 4.6 mm i.d.; mobile phase A = 1% *i*-PrOH–4% MeOH–95% water with 2.8 g of NaCl in 1 l of mobile phase A; mobile phase B = 10% *i*-PrOH–60% MeOH–30% water with 4.2 g of NaCl in 1 l of mobile-phase B; gradient, 0–5 min from 55% to 70% B, flow rate, 1.0 ml min^{-1}; KOT post-column reactor, 2.0 ml; ED, glassy carbon electrode at 0.8 V vs Ag/AgCl. Reprinted with permission from *Analytical Chemistry*, **62**, 2599. Copyright (1990) American Chemical Society

Dou and Krull (1990) recently applied HPLC–*hv*–ED to detect and determine aromatic amino acids, sulphur-containing amino acids, peptides containing such constituents and proteins. For example, Fig. 23 shows chromatograms for gradient elution of (a) standard insulin and (b) insulin formulation using on line *hv*-ED detection. The concentration of insulin in Fig. 23a was 100 μg ml^{-1} and concentrations as low as 10 μg ml^{-1} could be conveniently determined. In Fig 23c the chromatogram is for direct ED with the lamp off for the same insulin standard as in Fig. 23a. No insulin peak appeared without irradiation. The use of this detection technique in protein separations, reducing the size or changing the conformation of the biopolymers and generating more electrochemically active products than originally present, is advantageous in overcoming inherent limitations in the HPLC–ED of such biopolymers.

1.9 ALKANOLAMINES

The determination of alkanolamines is not always simple owing to the lack of an inherent chromophore or fluorophore. In addition, GC or HPLC can be difficult owing to the high polarity of these compounds. La Course *et al.* (1989) used ion-pair reversed-phase HPLC with on-line pulsed amperomatic detection at a gold electrode to separate mono-, di- and trialkanolamines.

1.10 AMIDES

The amide linkage is generally polarographically inactive unless it is incorporated into a ring system as with the barbiturate drugs. Anodic waves are observed at both the DME and solid electrodes (Kato and Dryhurst, 1973), but such electrochemical behaviour has not been recommended for the determination of these compounds at the trace level (Smyth *et al.*, 1975a).

Caprolactam (**XXII**) can be determined with an LOD of 0.2 μg ml^{-1} in waste waters and natural waters by AdSV after separation of the product of its reaction with *p*-(*N*,*N*-dimethylamino)benzene-*p*′-azobenzoyl chloride (Tocksteinova and Kopanica, 1987). The peak potential of the separated product was − 0.30 V.

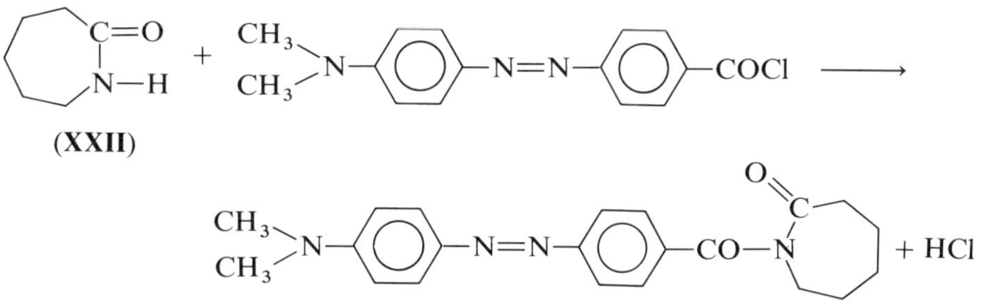

Phenytoin (**XXIII**), a cyclic amide, is an antiepileptic drug that has been determined by Eggers *et al.* (1982) at therapeutic levels of 10–20 μg ml^{-1} by homogeneous immunoassay using a drug labelled with the enzyme glucose 6-phosphate dehydrogenase. The immunoassay relies on a decrease in the rate of enzyme catalysis of $NAD^+ \rightarrow NADH$ when the antibody binds to the enzyme

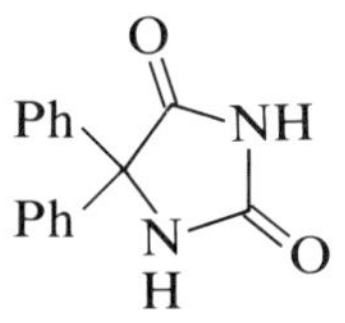

(XXIII)

labelled antigen as shown:

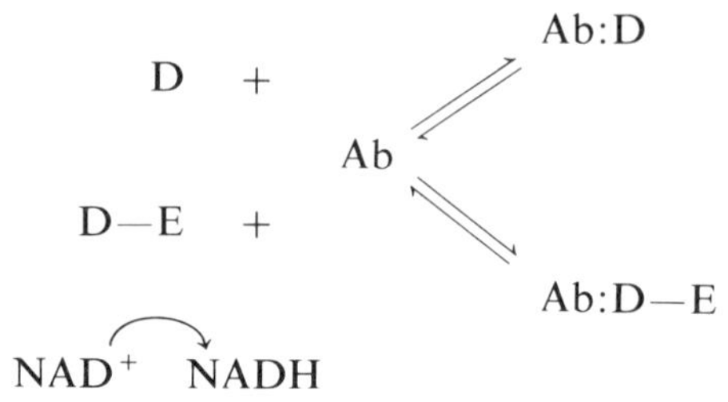

The catalytic activity of Ab:D—E is substantially diminished as a consequence of steric hindrance and limited substrate access. Since a larger concentration of D results in more D—E being in the unbound form and a greater turnover of NADH, the calibration plot is based on the rate of NADH formation vs concentration of D. This enzyme multiplied immunoassay technique (EMIT) is commonly used with spectrophotometric determination of NADH. Eggers *et al.* (1982) determined NADH by HPLC-ED at a glassy carbon electrode at + 0.75 V (vs Ag/AgCl) and were able, consequently, to determine phenytoin (**XXIII**). Antibody and proteins in the sample were removed by a reversed-phase C_{18} HPLC precolumn to prevent adsorption on the electrode and consequent fouling. Assays of serum samples from patients on phenytoin treatment gave results in good agreement with the standard spectrophotometric procedure at therapeutic levels.

1.11 ISOCYANATES

The properties of isocyanates to form polymers are widely exploited in industry, particularly in the manufacture of upholstery products, paint, varnishes, printing inks and adhesives. Organic isocyanates are, however, respiratory irritants and sensitisers and consequently a large number of industrial personnel are potentially at risk to occupational isocyanate exposure. Various HPLC methods have been published in recent years for the measurement of isocyanate vapours at trace levels in air. Of these, those methods with UV detection are the more established but recently fluorescence and electrochemical detection methods have also been reported. In all of these methods, the isocyanates are collected in 10 ml of a suitable derivatising solution contained in a bubbler or impinger in order to stabilise them prior to HPLC. 1-(2-Methoxyphenyl)piperazine was used by Warwick *et al.* (1981) to derivatise free monomeric aromatic and aliphatic isocyanates after their collection from a 10 l air sample. Using a 10 μl injection volume, a column packing of Hypersil ODS (150 mm × 4.5 mm i.d.), a mobile phase of acetate buffer (pH 6)-4 % acetonitrile and ED at + 0.80 V (vs Ag/AgCl), and LOD of 0.2 μg m^{-3} for the determination of phenyl isocyanate, toluene diisocyanate, hexamethylene diisocyanate and 4,4′-diisocyanate-diphenylmethane in air was calculated. Wu *et al.* (1990) used

Table 18. Dual detection of various polymeric isocyanate–tryptamine derivatives in HPLC

Substance	Type of isocyanate contained	Concentration of substance (ng ml^{-1})	Retention time of component (min)		Amounts of components in tryptamine equivalent (ng ml^{-1})	
			FD[a]	AD[b]	FD[a]	AD[b]
Mondur MT-40	TDI polymer	135	6.15	6.26	13	14
			7.29	7.41	45	48
			10.93	11.05	31	24
			17.05	17.33	10	7
Desmodur KL5-2444	HDI polymer	500	12.80	12.92	245	275
PMPPI[c]	PMPPI	107	11.01	11.13	57	63
			17.11	17.38	32	22
Mondur MR	MDI polymer	124	10.97	11.10	45	49
			17.04	17.28	33	34
			24.90	25.66	2.3	Trace
			29.20	29.77	8.6	9
Mondur CB-75	Aromatic polyisocyanate[d]	138	9.89	10.03	5.1	10
			11.95	12.15	4.2	6.8
			16.32	16.58	8.0	6.0
			20.20	20.54	24	20
			25.05	25.51	26	22

[a]Fluorescence detector.
[b]Amperometric oxidation detector.
[c]Not a commercial product; obtained from Aldrich.
[d]Precise identity unknown.

tryptamine (**XXIV**) as shown as an alternative derivatising agent prior to HPLC with dual detectors based on fluorescence emission and amperometric oxidation Table 18. Meyer and Tallman (1983) used *p*-aminophenol as the derivatising agent. The detector in this latter case was Kel-F graphite composite electrode which gave an LOD of 94 pg of toluene diisocyanate injected, a linear working range beyond 12 ng of toluene diisocyanate injected and a relative standard deviation of < 2%.

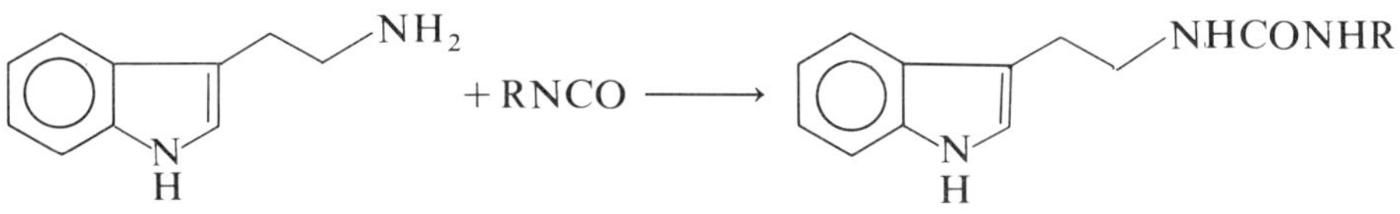

(**XXIV**)

2

Voltammetric Analysis of Selected Sulphur-containing Molecules

2.1 DISULPHIDES AND DITHIOCARBAMATES

The disulphide linkage exists in many naturally occurring molecules and is directly reducible at the DME in accordance with the overall reaction

$$\mathrm{R{-}S{-}S{-}R} + 2e^- + 2\,\mathrm{H}^+ \rightarrow 2\,\mathrm{R{-}SH}$$

Polarographic methods have been used to determine disulphides such as oxidised glutathione (Mairesse-Ducarmois *et al.*, 1975), lipoic acid (Mairesse-Ducarmois *et al.*, 1977), vasopressin and the fungicidal agent thiram [tetramethylthiuram disulphide; $(CH_3)_2NC({=}S)SSC({=}S)N(CH_3)_2$]. Thiram can be determined in plants down to a detection limit of 0.1 ppm using reductive polarography (Nangniot *et al.*, 1978) or down to 10^{-8} M in pure solutions using CSV at a mercury-plated platinum electrode, after reduction of the —S—S— functional group to an —SH group by ascorbic acid (Brand and Fleet, 1970). The related disulphide disulphiram (tetraethylthiuram disulphide) can be directly determined by DPP down to 5×10^{-7} M (Mairesse-Ducarmois *et al.*, 1976) or by determining the CS_2 breakdown product of the disulphide by complexation with Cu(II) (Brown *et al.*, 1974). Dithiocarbamates such as ethylene-1,2-bisdithiocarbamate, [$HS(C{=}S)NH(CH_2)_2NH(C{=}S)SH$] can be determined by anodic polarography with a detection limit of the order of 10^{-6} M or by CSV with a detection limit of 10^{-7} M after pre-electrolysis for 2 min at 0 V (vs SCE) at a mercury-plated platinum electrode (Brand and Fleet, 1968).

Metal-containing dithiocarbamate pesticides can either be determined by voltammetry of the intact molecule [e.g. zineb (zinc ethylene bisdithiocarbamate) has an LOD of 1 ppm in 0.1 M NaOH supporting electrolyte using linear-sweep voltammetry (Nangniot, 1960)] or by its destruction followed by ASV of the liberated metal ion {e.g. Sauberlich *et al.* (1969) determined ferbam [iron(III) dimethyldithiocarbamate] down to an LOD of 0.5 ng ml^{-1} in pure solution} or by anodic polarography of the resulting dithiocarbamate anion (Halls *et al.*, 1968).

2.2 THIOBARBITURATES

Smyth *et al.* (1970a,b, 1971) investigated extensively the d.c. polarographic behaviour of a wide range of thiobarbiturates. These molecules give rise to both anodic and cathodic waves at the DME with the mechanisms of reduction and oxidation being greatly dependent on the pH and nature of substitution in the molecule. The cathodic waves in 0.05 M borate buffer (pH 9.3) were recommended for analytical purposes. Smyth *et al.* (1975a) reported LODs of the order of 10^{-7} M in Britton–Robinson buffer (pH 8.0) for the DPP determination of thiobarbiturates.

2.3 THIOUREAS

Thioureas give rise to anodic phenomena at mercury electrodes owing to the formation of partially insoluble mercury–thiourea compounds/complexes. This can result in LODs of 10^{-7} M by DPP of the anodic process or 1 ppb by CSV (Fig. 24) (Smyth and Osteryoung, 1977). A polarographic method was applied to the determination of 1,3-ethylenethiourea (**XXV**) in food by Engst and Schnaak (1974). This method involved extraction with chloroform–ethanol (9 + 1), clean-up on an alumina column, separation by paper chromatography

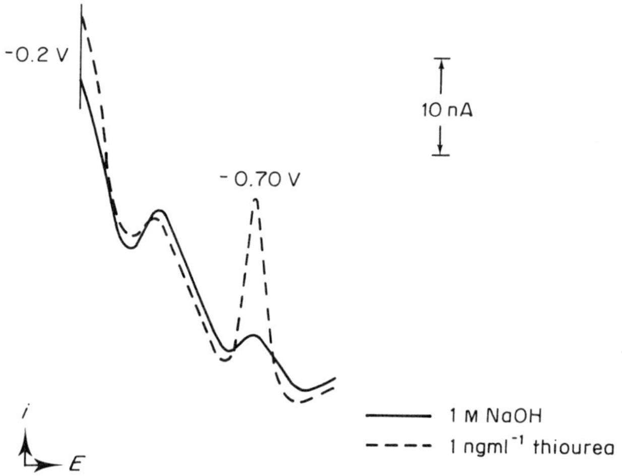

Figure 24. DPSCV of 1 ng ml^{-1} thiourea in 1 M NaOH (Smyth and Osteryoung, 1977) Solid line, 1 M NaOH; dashed line, 1 ng ml^{-1} thiourea. Reprinted with permission from *Analytical Chemistry*, **49**, 2310. Copyright (1977) American Chemical Society

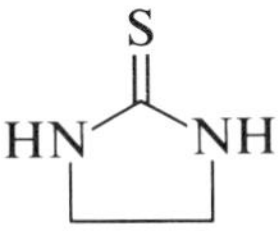

(**XXV**)

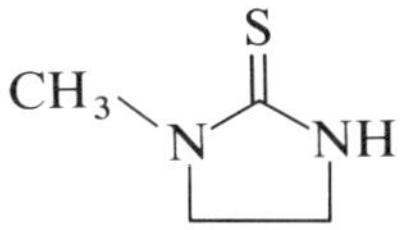

(**XXVI**)

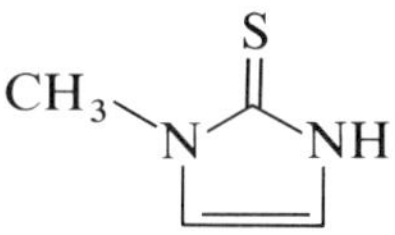

(**XXVII**)

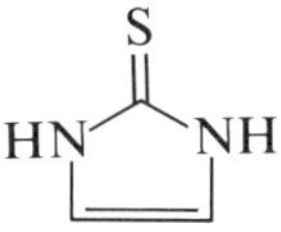

(**XXVIII**)

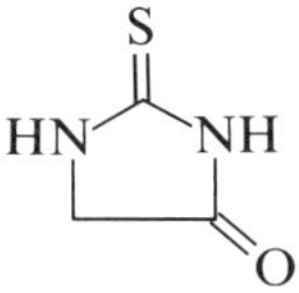

(**XXIX**)

and nitrosation to give a reducible entity. The recovery was approximately 70% and the method used to determine concentrations of (**XXV**) in the range 0.05–2 μg ml^{-1} in potatoes, tomatoes, meat and liver.

Lawrence *et al.* (1981) used HPLC with either UV or polarographic detection to determine 1,3-ethylenethiourea (**XXV**) and related sulphur compounds *N*-methylethylene thiourea (**XXVI**), methimazole (**XXVII**), thioimidazole (**XXVIII**) and thiohydantoin (**XXIX**). Nanogram amounts of these molecules can be satisfactorily resolved using normal-phase HPLC with UV detection or reversed-phase HPLC with polarographic detection. Rat urine could be directly subjected to the latter technique with detection at + 0.2 V to detect simultaneously (**XXV**) and (**XXVIII**) (Fig. 25).

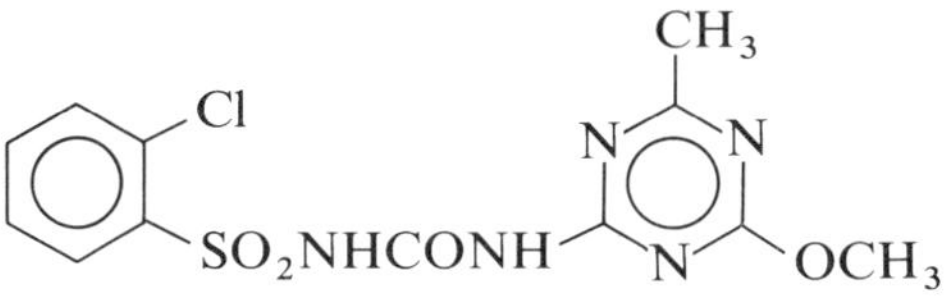

(**XXX**)

2.4 SULPHONYLUREAS

Concialini *et al.* (1989) studied the DPP behaviour of selected sulphonylurea herbicides, characterised by low crop application rates, excellent crop selectivity and low mammalian toxicity. They found LODs of the order of 10^{-8} M for actives such as chlorsulphuron (**XXX**).

2.5 THIOAMIDES

The d.c. polarographic determination of ethionamide in urine samples has been reported by Okuda (1963), who was also able to follow polarographically the

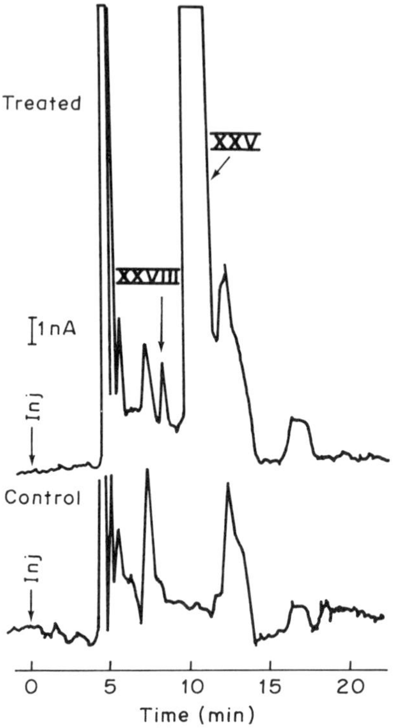

Figure 25. Reversed-phase HPLC with polarographic detection of rat urine sample. $E = +200$ mV. Reproduced by permission of Elsevier Science Publishers from *J. Chromatogr.*, **212** (1981) 245.

metabolism of this molecule to its amide and carboxylic acid metabolites. Bieder and Brunel (1971) used a.c. polarography to resolve ethionamide and seven of its metabolites. Davidson and Smyth (1977) described a DPCSV method of analysis which can directly determine thioamides with functional groups $RC(=S)NH_2$ in plasma and urine down to a concentration of 2×10^{-8} M. Naturally occurring substances such as glutathione, thiamine, methionine, cystine and cysteine and anionic species such as Cl^- were shown not to interfere with the determination of compound **XXXI**, as evidenced by Fig. 26. It appears that at the deposition potential, the thioamide reacts with Hg(II) to give HgS, which then adsorbs on the Hg surface and is subsequently stripped from the surface in the cathodic scan. Thioamide metabolites which do not contain the sulphur entity, such as the amide and carboxylic acid, do not undergo this reaction, thus endowing the method with an even greater selectivity.

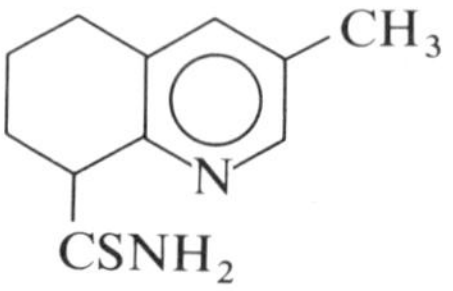

(**XXXI**)

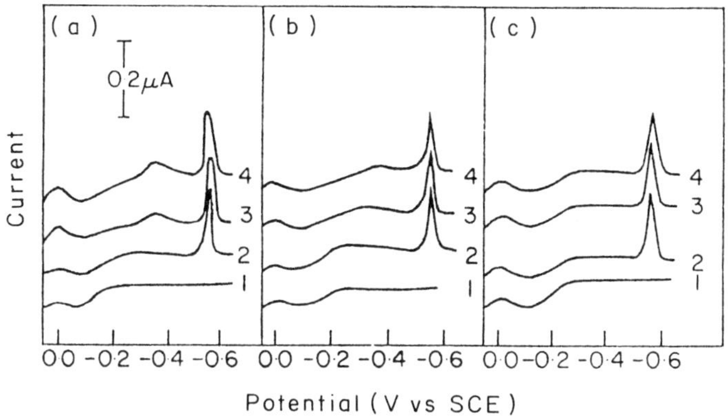

Figure 26. Effect of possible interferents on stripping voltammograms for 1:1 horse plasma: pH 4.78 buffer solutions, 3.7×10^{-7} M solutions of compound **XXXI**. (a) Additions of (2) 0.0, (3) 12.0 and (4) 24.0 μg of cysteine; (b) additions of (2) 0.0, (3) 3.0 and (4) 6.0 μg of reduced glutathione; (c) additions of (2) 0.0, (3) 15.0 and (4) 30.0 mg of sodium chloride. Scans (1) in each case are for blanks containing no drugs or additives. Reprinted with permission from *Analytical Chemistry*, **49**, 1195. Copyright (1977) American Chemical Society

2.6 ISOTHIOCYANATES

Moreira *et al.* (1989) investigated the AdSV of isothiocyanates. Fluorescein isothiocyanate in phosphate buffer (pH 8) gives a single adsorptive stripping voltammetric peak at an HMDE at − 0.88 V (vs Ag/AgCl) after adsorption at 0 V. The peak was found to be due to the reduction of the fluorescein moiety in the isothiocyanate. When accumulation was carried out at potentials more positive than 0 V, a second peak was observed at − 0.60 V due to HgS formation during the accumulation period. Two derivatives of isothiocyanates with amino acids, phenylthiohydantoin-tyrosine and methylthiohydantoin-glycine, were shown to give the − 0.60 V peak when accumulation was effected at 0 V from phosphate buffer (pH 8.0). Calibration graphs were linear in the range 2×10^{-7}–2×10^{-8} M.

2.7 THIOLS

Thiols, in common with many of the sulphur-containing organic molecules discussed, produce anodic waves at the DME owing to the formation of partially insoluble mercury compounds/complexes and are thus amenable to CSV methods of analysis, e.g. 2-mercaptopyridine 1-oxide can be determined down to 8×10^{-10} M (Csejka *et al.*, 1975), pyridine-2-thiol down to 8×10^{-9} M (Fernandez Alvarez and Smyth, 1989) cysteine down to 10^{-9} M (Grier and

Andrews, 1981) and ethyl xanthate to 10^{-8} M (Ivaska and Leppinen, 1986). Thiols (and disulphides) can be simultaneously determined using HPLC and thin-layer dual mercury amalgam electrodes in the series mode (Allison and Shoup, 1983). After HPLC separation, disulphides were first converted into the corresponding thiols at an upstream electrode held at − 1.00 V (vs Ag/AgCl). Both thiols and disulphides were then detected as thiols downstream at +0.15 V due to oxidation at a mercury surface. The selectivity of this reaction gives rise to a low level of interference and has been applied to the determination of oxidised and reduced glutathione in human blood after a simple dilution. The method was found to be much more rapid than corresponding enzymatic methods.

Halbert and Baldwin (1985) determined cysteine and glutathione in plasma and blood by HPLC–ED using a chemically modified carbon paste electrode incorporating cobalt phthalocyanine with LODs of < 4 pmol. The electrocatalytic activity of these chemically modified electrodes permitted optimum response at + 0.75 V (vs Ag/AgCl), several hundred mV lower than that required at conventional carbon electrodes. Wring *et al.* (1989) used a similar detector for the determination of glutathione (GSH) in human plasma (Fig. 27). The mean circulating level of GSH in plasma, found in three normal subjects, was 2.69 μM, in agreement with previously reported values. The LOD was found to be 0.17 μM.

Electroinactive penicillins can be converted into the corresponding penicilloic acids by alkaline hydrolysis followed by accumulation at an HMDE at − 0.10 V (vs SCE) in a pH 4.6 buffer containing an excess of Cu(II). During the accumulation, the penicilloic acid is further degraded and a copper complex of penicillamine $[(CH_3)_2C(SH)CH(COOH)NH_2]$ is formed. This complex is subsequently reduced at − 0.40 V in the cathodic stripping step. After accumulation for 10 min 2×10^{-10} M penicillin can be detected (Forsman, 1983).

D-Penicillamine is used for the treatment of Morbus Wilson disease, metal intoxication, cystinuria and rheumatoid arthritis and can be determined in plasma and urine by separation on a cation-exchange resin and oxidative ED of the SH group at a gold indicator electrode (Kreuzig and Frank, 1981). The method has been automated and separations from plasma and urine take 7 and 9 min, respectively. The LODs are 0.05 $\mu g\ ml^{-1}$ in plasma and 0.20 $\mu g\ ml^{-1}$ in urine with a relative standard deviation of 2.9% ($n = 10$). Drummer *et al.* (1986) improved on this method by use of a high-efficiency C_{18} reversed-phase column and were able to determine *N*-acetylcysteine, and also penicillamine, with an LOD of 25 ng ml^{-1} using a 500 μl blood sample.

Zhu and Brajter-Toth (1990) electrochemically oxidised the antineoplastic drug 6-thiopurine (**XXXII**) in a high-pressure coulometric cell of 5 μl volume. This cell is in an on-line configuration with HPLC, which separates the oxidation products prior to their determination using UV detection. In the potential range 0.4 to 0.8 V (vs Pd), intermediate purine-6-sulphenic acid (RSOH) could be detected together with purine-6-sulphinic acid ($RSO_2^-H^+$) and 6-thiopurine disulphide (RSSR). At potentials >0.8 V, purine-6-sulphonic

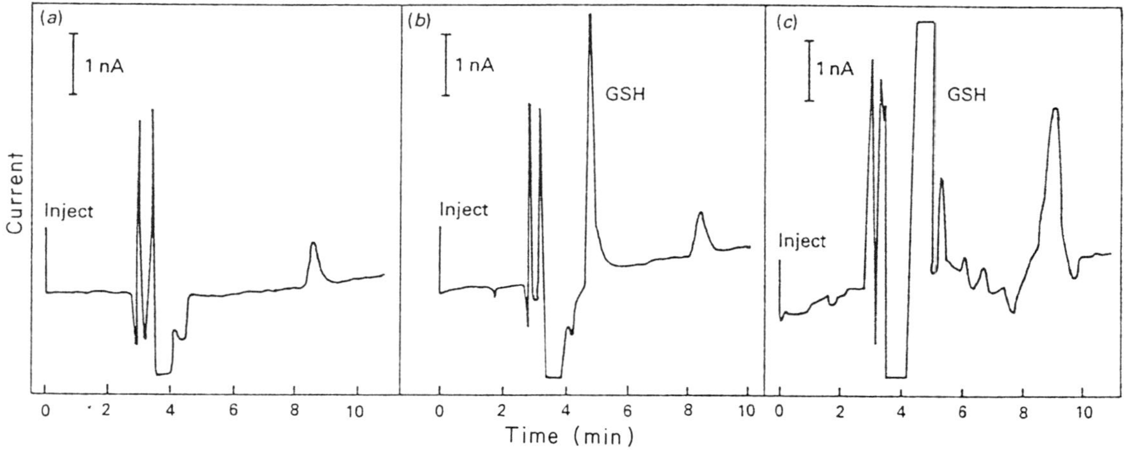

Figure 27. Chromatograms obtained by HPLC–ED of (a) GSH-free control serum, (b) control serum spiked with 1.54 μg ml^{-1} of GSH (final concentration 5 μM) and (c) a typical normal human plasma (final concentration 1.87 μM). Reproduced by permission of the Royal Society of Chemistry

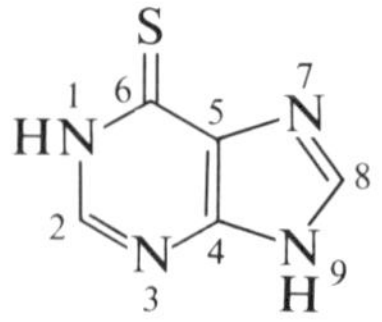

(XXXII)

acid ($RSO_3^-H^+$) was detected and the oxidation of 6-thiopurine was completed. Intermediates and products formed in the horseradish peroxidase-catalysed oxidation of 6-thiopurine were also studied. Enzymatic activation with horseradish peroxidase was similar to electrochemical oxidation at <0.8 V.

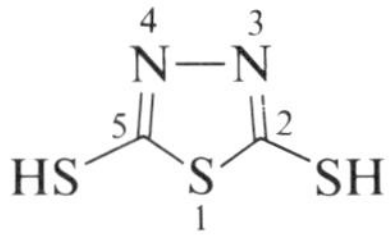

(XXXIII)

Chiang *et al.* (1988) studied various sulphur-containing ligands for the AdSV of Cd and Zn and concluded that 2,5-dimercapto-1,3,4-thiadiazole **(XXXIII)** (DMTD) was the preferred ligand. They developed a method for the direct determination of these metals in biological samples using benzyltrimethylammonium methoxide as the digesting solvent. For example, the certificate for an NBS spinach sample lists three major constituents, K, Ca and Mg, and thirteen trace metal constituents of which Al (870 $\mu g\ g^{-1}$), Fe (550 $\mu g\ g^{-1}$), Mn (165 $\mu g\ g^{-1}$) are present in the largest amounts. Ten separate determinations for Zn were carried out (sample sizes ranging between 195 and 291 mg) and yielded an average value of $49.10 \pm 0.01\ \mu g\ g^{-1}$. This compared well with the certified value of $50 \pm 2\ \mu g\ g^{-1}$. Arsenic has also been measured in oyster tissue (Chiang *et al.*, 1989) by AdSV after digestion of the sample with benzyltrimethylammonium methoxide and chelation with **XXXIII** (DMTD). An attempt was also made to extend the method to the simultaneous determination of As, Se and Ni in oyster tissue but this resulted in some distortion of both peak potential and peak shape (Fig 28).

2.8 SULPHONAMIDES

Sulphonamides have been available as chemotherapeutic agents since they were patented in 1935. Sulphadiazine **(XXXIV)** is prepared as follows:

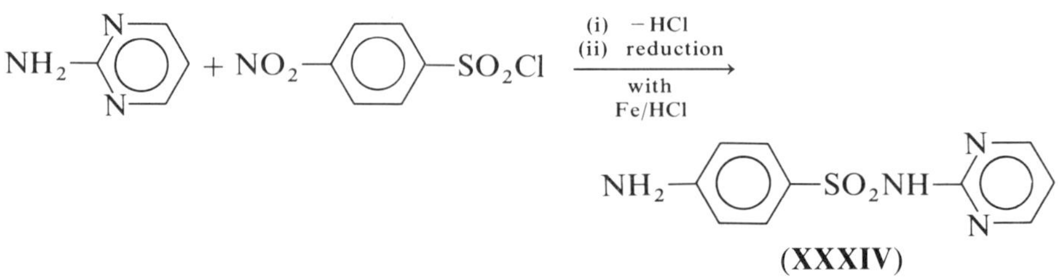

(XXXIV)

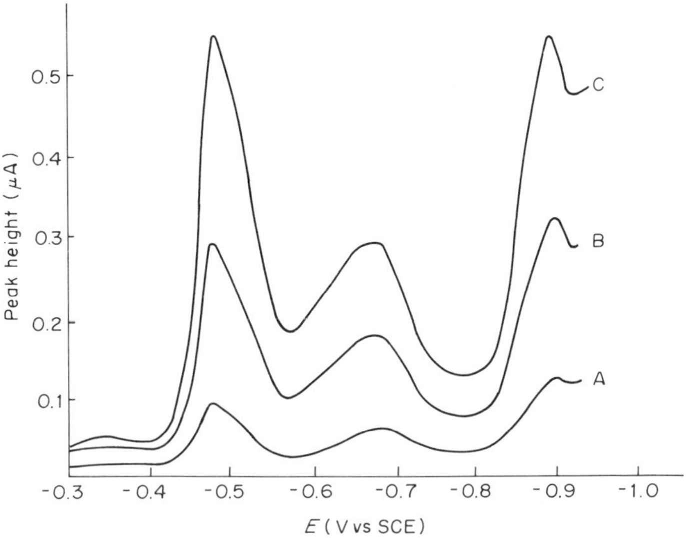

Figure 28. Voltammograms for the simultaneous determination of arsenic, selenium and nickel in Oyster tissue (sample mass = 297 mg). A = sample; B = 10 ng ml^{-1} As + 1.5 ng ml^{-1} Se + 1.0 ng ml^{-1} Ni; C = 20 ng ml^{-1} As + 3.0 ng ml^{-1} Se + 2.0 ng ml^{-1} Ni. Reproduced from *Mikrochim. Acta*, 1989, **II**, 149, by permission of Springer-Verlag, Vienna

Direct spectrophotometric methods have rarely been used to determine **XXXIV** in raw materials and formulations. More commonly, a derivative of the sulphonamide is prepared, such as by diazotisation of the primary aromatic amine grouping, followed by a coupling reaction with Bratton–Marshall reagent and determination of the product by visible spectrophotometry. Such a derivatisation is selective if the potential interferences such as process impurities and degradation products do not contain a primary aromatic amine group. For example, the starting material, 2-aminopyrimidine, would be expected to undergo this derivatisation reaction. The same can be said about a similar derivatisation reaction followed by a polarographic finish (Fogg and Ahmed, 1974). Sulphonamides such as **XXXIV** can be oxidised at a glassy carbon electrode through a reaction involving the primary aromatic amine group, and this has been used for quantitative analysis by Momberg *et al.* (1984)—again, any process impurity or degradation product containing this functional group will interfere. DPP at the DME could be of value in the determination of potential process impurities containing the aromatic NO_2 group, which are reduced at small negative potentials.

Chromatographic methods, in particular HPLC, show superiority in terms of selectivity for the determination of sulphonamides and the directly electroreducible trimethoprim (in trimethoprim-potentiated sulphonamide formulations). Sulphadiazine (**XXXIV**) and its formulated product, sulphadiazine tablets, are now officially assayed according to the US Pharmacopeia (XXI Revision) by

HPLC with a stationary phase of octadecylsilane chemically bonded to porous silica microparticles, 5–10 μm in diameter, and a mobile phase of degassed water–acetonitrile–glacial acetic acid (87 + 12 + 1) with UV detection at 254 nm. The US Pharmacopeia also states HPLC to be its official method of assay for sulphamethoxazole and trimethoprim oral suspensions and tablets.

2.9 SULPHONATES

Sulphonate compounds can only be reduced in aprotic media and the waves obtained are not recommended for trace analytical purposes. They also give rise to tensammetric waves in aqueous solutions (caused by desorption of the species at the mercury surface, thus effecting a change in the double-layer structure) which are manifested as sharp incisions in a.c. polarography and can be used quantitatively down to a concentration of ca 10^{-6} M.

Sulphonates can also be determined indirectly by their complexation with dyes such as methylene blue (Buchanan and Griffith, 1963) or by nitration if they contain a phenyl moiety (Hart *et al.*, 1976). A polarographic method has also been described for the determination of sodium dodecyl sulphate and sodium dodecylbenzene sulphonate in surface water (Shinozuka *et al.*, 1972). It is necessary to contrast these polarographic methods for alkylbenzene sulphonates with superior chromatographic methods such as HPLC–UV detection with 5–200 ng determinable (Matsueda *et al.*, 1982), HPLC–fluorescence detection (LOD for S/N = 3 was found to be 1.5 ng per component injected (Castles *et al.*, 1989) and GLC–MS for the determination of ppb concentrations (Hon-Nami and Hanya, 1978). DPP has also been used for the determination of lignin sulphonic acids in natural waters (Eberle *et al.*, 1969).

2.10 SULPHIDES AND SULPHOXIDES

Johansson (1983) has made a detailed investigation of the electrochemical reduction of selected sulphides and sulphoxides of pharmacological importance as gastric secretion inhibitors. Their general structure can be represented as heterocycle′—HCR—S—heterocycle″, where R=H or methyl and the S atom

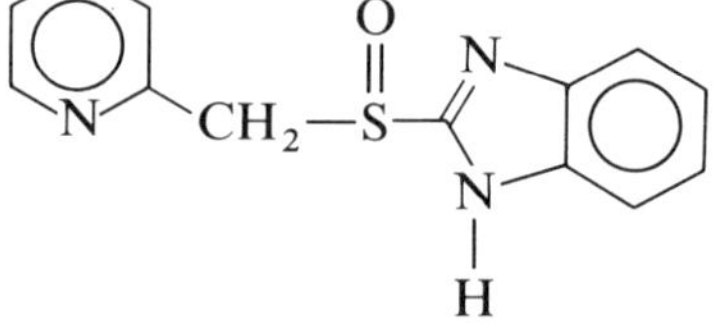

(XXXV)

can be oxidised to produce a sulphoxide. Heterocycle′ is pyridine or 3-methylpyridine and heterocycle″ is a benzimidazole, a methyl-substituted benzimidazole or an imidazole. Most compounds are irreversibly reduced at the DME by cleavage of the CH_2—S bond. The sulphoxides consumed $4e^-$ and the sulphides $2e^-$ per molecule in the reduction process. DPP was used in the determination of 2-benzimidazolyl-2-pyridylmethyl sulphoxide (**XXXV**) in pharmaceutical formulations and the results agreed with those obtained by HPLC. The relative standard deviation was $< 1\%$ in the DPP determination (Johansson and Persson, 1978).

3

Voltammetric Analysis of Selected Oxygen-containing Molecules

3.1 ALDEHYDES AND KETONES

Aliphatic aldehydes are reduced according to

$$\mathrm{RCH(OH)_2} \underset{+\mathrm{H_2O}}{\overset{-\mathrm{H_2O}}{\rightleftharpoons}} \mathrm{RCHO} \overset{+1e^- +1H^+}{\rightleftharpoons} \mathrm{R\dot{C}HOH} \xrightarrow{+1e^- +1H^+}$$

$$\mathrm{R\bar{C}HOH} \xrightarrow{\mathrm{H_2O}} \mathrm{RCH_2OH + OH^-}$$

and the process is governed by the rate at which dehydration takes place. There are direct polarographic methods for the determination of simple aliphatic aldehydes, such as formaldehyde, acetaldehyde and butyraldehyde, in waste waters with LODs in the low μg ml^{-1} region (Kuchumova *et al.*, 1963). The reversibility exhibited by aliphatic carbonyl compounds on the first electron-consuming step results in an enhancement of the waves obtained for these compounds during a.c. polarography. Fleet *et al.* (1973) utilised this effect for the determination of mesityl oxide in the presence of diacetone alcohol, using a fast-scan, phase-sensitive a.c. technique.

Aromatic aldehydes and ketones are irreversibly reduced at the DME in either one- or two-electron processes (Zuman and Perrin, 1969). Well defined waves/peaks of analytical utility have been obtained for several compounds of biological importance. For example, substituted benzophenones are commonly encountered in the analytical chemistry of the 1,4-benzodiazepines (Clifford and Smyth, 1974) and, in the case of bromazepam, de Silva *et al.* (1974b) used DPP in order to determine urinary benzophenone metabolites of this benzodiazepine.

Tetracyline antibiotics possess a complicated reduction mechanism which partly involves reduction of the keto group in ring A of the molecule (Caplis, 1970). Yarnitzky and Smyth (1991) investigated the square-wave voltammetric behaviour of oxytetracycline at the SMDE and showed that it can be determined by AdSV down to 10^{-8} M and that oxytetracycline can be resolved from the antibiotics and antibacterials chloramphenicol, cephalothin and trimethoprim in a single scan. Wang *et al.* (1986) achieved LODs of 10^{-9}–10^{-10} M for a series of tetracyclines using 300 s accumulation prior to cathodic stripping. Hou and Wang (1989) developed an HPLC–ED method for determination of tetracycline antibiotics with LODs of 0.2 ng for oxytetracycline and tetracycline, 1.2 ng for chlortetracycline and 4 ng for doxycycline.

Telting-Diaz *et al.* (1990) studied the electrochemical behaviour of the antimicrobial agent pipemidic acid {8-ethyl-5,8-dihydro-5-oxo-2-(1-piperazinyl)-pyrido[2,3-*d*]pyrimidine-6-carboxylic acid} (**XXXVI**) at the HMDE and then developed an AdSV method for its determination in urine samples. Voltammetric determination was carried out at – 0.76 V in 0.1 M $HClO_4$, a reduction process possibly corresponding to the keto group in **XXXVI**. AdSV has also been applied (Wang and Mahmoud, 1986) to the determination of the antibiotics streptomycin and erythromycin down to 10^{-9}–10^{-10} M concentrations. Voltammetric determination was carried out at – 1.58 and – 1.2 V, the reduction processes probably corresponding to the aliphatic aldehyde and ketone moieties, respectively.

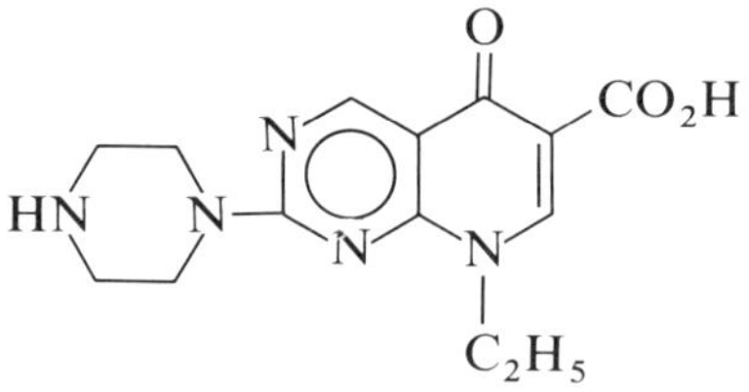

(**XXXVI**)

Derivatisation has been widely employed for the polarographic or voltammetric determination of aldehydes and ketones. Afghan *et al.* (1975) used semicarbazone formation at pH 4–6 followed by determination in citrate buffer–EDTA by twin-cell potential-sweep voltammetry with an LOD of 0.25 ppb. This LOD was claimed for the determination of such carbonyl compounds in natural waters and industrial effluents without any separation or preconcentration of the sample. The method, however, could not differentiate between many aldehydes and ketones. Thus for improved selectivity, it is necessary to introduce a separation step, e.g. HPLC, as was demonstrated by Vigh *et al.* (1980) for 2,4-dinitrophenylhydrazone derivatives and by Creech *et al.* (1982), who trapped aldehydes and ketones from diesel exhaust in 2,4-dinitrophenylhydrazine solution prior to HPLC–UV detection with LODs > 0.01 ppm. Liu and Abruna (1989) carried out the derivatisation step for aromatic aldehydes in carbon paste electrodes, modified with aromatic primary amines such as $[Fe(CN)_5L]^{3-}$, where L = 4-aminopyridine, to yield imines whose redox response was used as the analytical signal. This resulted in determination of benzaldehyde at sub-μM levels with no detrimental effects on the analytical signal from aromatic ketones such as benzophenone.

Hart (1986) derivatised the quinone vitamin K_1 to its hydroquinone after HPLC separation at an upstream electrode held at -1.3 V followed by detection of the hydroquinone at a downstream electrode at 0 V. This was used for the measurement of depressed circulating levels of vitamin K_1 in patients with osteoporosis. These levels were of the order of 60 pg ml^{-1} in plasma and the LOD was about 30 times lower than that given by HPLC with UV detection. Solvent extraction–AdSV gave an inferior LOD of 450 ng ml^{-1} in plasma (Hart *et al.*, 1989). Vire *et al.* (1988) applied square-wave AdSV to the determination of menadione (vitamin K_3) over the concentration range 5×10^{-7}–2×10^{-10} M with an LOD of 1.3×10^{-10} M. Stripping voltammetry was applied after preconcentration of the reduced compound (the hydroquinone) and scanning the potential towards less negative values.

La Course *et al.* (1985) separated intact aldehydes and ketones by HPLC prior to detection and determination by a photoelectrochemical derivatisation technique. They designed a flow-through, thin-layer amperometric cell modified to irradiate the working electrode surface and so take advantage of new and/or altered electrochemical properties of photogenerated excited states, intermediates or products. Such photoelectrochemical detection has been found to be responsive to alkyl and aryl ketones and aldehydes, i.e. to compounds that undergo n–π^* excitation as opposed to π–π^* excitation. The response was linear over 3–4 orders of magnitude and LODs for conjugated carbonyl compounds were in the range 2–10 ng, comparable to those with UV detection. Figure 29 shows the HPLC–photoelectrochemical detection of a mixture of carbonyl compounds with and without irradiation.

Kuwahara *et al.* (1990) determined the 5-lipoxygenase inhibitor CGS 8515 (**XXXVII**) in plasma by HPLC–ED, detection being accomplished by reduction of the naphthoquinone ring at a relatively low negative potential of -0.15 V. Less interference from endogeneous substances was obtained compared with oxidative detection, utilising the secondary aromatic amine group. At the low potential of detection, i.e. -0.15 V, dissolved oxygen was also not a significant

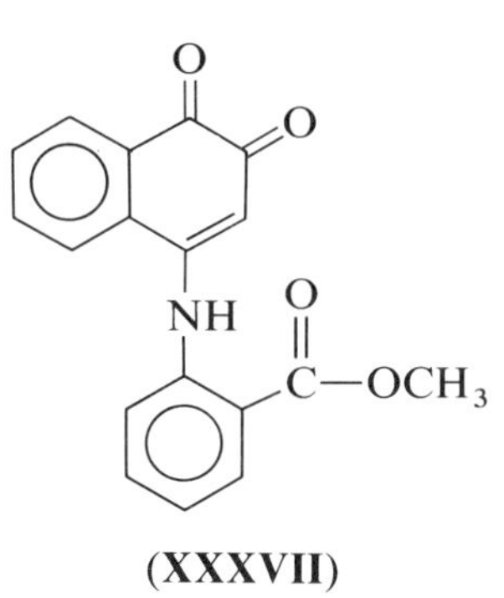

(**XXXVII**)

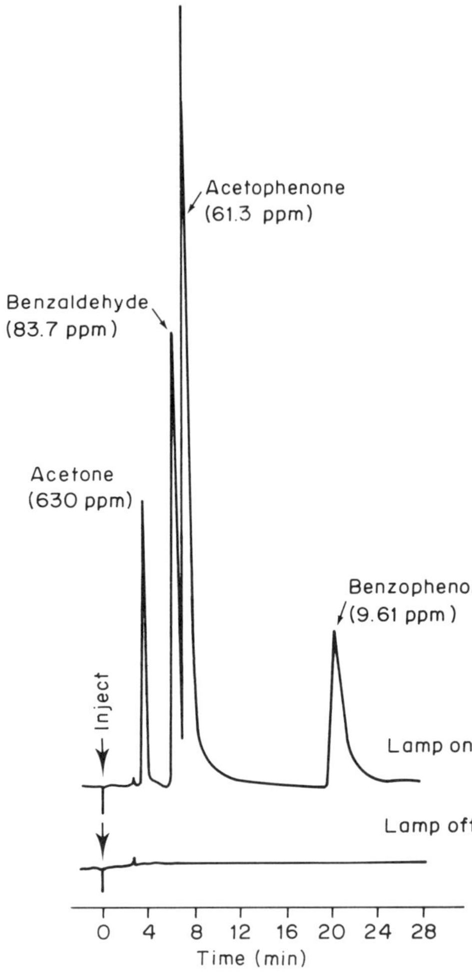

Figure 29. HPLC-*hν*-ED of a mixture of carbonyl compounds with and without irradiation. Mobile phase, methanol-0.2 M sodium chloride (1 + 1) at 0.8 ml min^{-1}; 200 μl injection: glassy carbon electrode, 0.00 V (oxidative) and 500 nA V^{-1}. Reprinted with permission from *Analytical Chemistry*, **57**, 1810. Copyright (1985) American Chemical Society

problem in the resulting assay. Plasma samples from orally treated rat, dog and monkey were determined for **XXXVII** as shown in Fig. 30. The rat showed two apparent metabolites at 5.8 and 13.7 min, the dog showed one at 5.2 min and the monkey showed four at 4.2, 5.7, 7.4 and 13.0 min.

3.2 PHENOLS

Phenols are widely distributed as molecules of biological significance in that they are found as drugs, pollutants, vitamins, plant metabolites, growth pro-

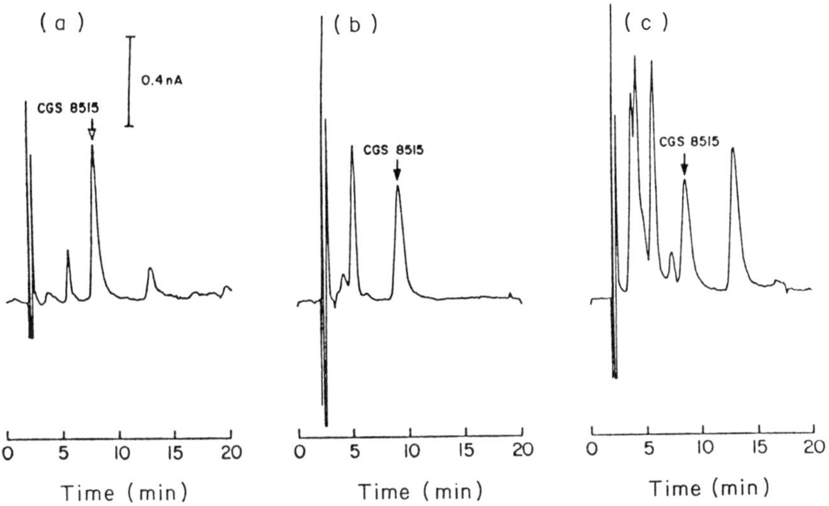

Figure 30. Chromatograms of plasma extracts after the oral administration of **XXXVII** from (a) rat (100 mg kg^{-1}; 2 h after dosing), (b) dog (250 mg kg^{-1}, 2 h after dosing) and (c) cyanomolgus monkey (127 mg kg^{-1}; 6 h after dosing). Reproduced by permission of Elsevier Science Publishers from *J. Chromatogr.*, **534** (1990) 260

moters and neurochemically active molecules. Such phenols can be conveniently oxidised on-line at solid electrodes following HPLC to produce analytical methods of choice over GC, which requires prior derivatisation, and HPLC–UV detection, with generally inferior detection limits. A selection of appropriate applications is given in Table 19. Figure 31 shows the superiority of ED over UV detection after HPLC separation of selected phenols in water samples (Armentrout *et al.*, 1979) and Table 20 shows the superior LODs for amperometric detection of the phenol narciclasine (1,2,3,7-tetrahydroxy-8,9-methylenedioxy-1,2,3,4-tetrahydrophenanthridone) at a carbon fibre array working electrode compared with fluorescence detection and UV detection following reversed-phase HPLC (Svagrova *et al.*, 1991).

Olefirowicz and Ewing (1990) were able to measure dopamine concentrations in the cytoplasmic compartment of single neurons of *Planorbis corneus* by capillary electrophoresis with off-column ED set at a potential of + 0.7 V (vs SCE) (Fig. 32). The peaks in this electropherogram were identified by calculating the electrophoretic mobilities of each peak and comparison with authentic standards. A small increase in migration time observed for the cytoplasmic sample relative to those of authentic standards is thought to be due to increased viscosity in the cytoplasmic sample relative to that of the standards.

Preconcentration and determination of the cancer chemotherapy drug doxorubicin (adriamycin) have been achieved by a flow-injection approach utilising adsorption of the drug on a carbon paste electrode, medium exchange and differential-pulse voltammetry on the adsorbing surface (Chaney and Baldwin,

Table 19. HPLC–ED of phenols of biological significance

Type	Molecules	Notes	Ref.
Drugs	Ciramadol, dezocine	Determined in body fluids after simple one-step extraction	Locniskar and Greenblatt (1986)
	SK&F 89124, SK&F 85738	Determined in rat, dog and human plasma. Sample preparation with liquid–solid extraction prior to HPLC–ED at + 0.74 V (vs Ag/AgCl)	Swagzdis *et al.* (1985)
	Morphine, dilaudid, naltrexone, naloxone	The four opiates are separated in a single HPLC run of 15 min with ED at + 0.85 V	Connor *et al.* (1989)
	Isoxsuprine	Method used to study pharmacokinetics of orally administered isoxsuprine in horses. LOD 1 ng ml^{-1} in plasma	Hashem and Lubczyk (1991)
Pollutants	Chlorophenols	HPLC performed with a polymeric cation-exchange resin column, acidic acetonitrile–water eluent and an electrochemical detector of carbon black/polyethylene tubular anode. Detection limits 0.2 ppb (phenol) to 0.7 ppb (pentachlorophenol)	Armentrout *et al.* (1979)
	Nitrophenols, chlorophenols	Detector uses parallel–adjacent glassy carbon electrodes held at − 0.75 V (for nitrophenols) at + 0.98 V (vs Ag/AgCl) (for all oxidisable phenols)	Roston *et al.* (1982)
	Phenol, hydroxyphenols, alkylphenols, chlorophenols	Glassy carbon dual electrode provides LODs ranging from 55 pg for hydroquinone to 19 ng for pentachlorophenol (injection volume for 20 μl)	Hagen *et al.* (1991)
Vitamins	Vitamin E	Extraction of 25 μl of serum with 100 μl of ethanol and 100 μl of heptane. Heptane phase evaporated and residue dissolved in methanol prior to reversed-phase HPLC with thin-layer cell for electrochemical detection at + 1.0 V. Normal circulating levels measured, i.e. 4.3–9.7 $\mu g\ ml^{-1}$ (α-), 1.8–3.9 $\mu g\ ml^{-1}$ (β- and γ-)	Chou *et al.* (1985)

	Vitamin B_6 compounds (i.e. 3-hydroxy pyridines)	Pyridoxal and pyridoxal phosphate gave maximum currents at + 1.0 V and onset of oxidation did not occur until +0.7 V. Cell was used in oxidative screen mode with the upstream electrode at + 0.6 V and the downstream electrode at + 1.0 V. HPLC–ED gave a linear response in the range 200 pg–100 ng	Hart (1986)
Plant metabolites		Water–ethanol extracts from plant material subjected to HPLC–ED. Single-electrode electrochemical detection 2–5 times more selective than UV detection; dual-electrode parallel ED can reduce LODs by a factor up to 5 for phenolics with high oxidation potentials	Nagels and Creten (1985)
Growth promoters		HPLC–ED applied to determination of growth-promoting hormones in meat. LODs are oestrone and hexoestrol 1–2 ng g^{-1}, 17 β-oestradiol, diethylstilboestrol and dienoestrol 2–3 ng g^{-1}	Smyth and Frischkorn (1980)
Neurochemicals	Norepinephrine, 3,4-dihydroxy-phenylacetic acid, dopamine, 5-hydroxyindole-3-acetic acid, homovanillic acid, serotonin	Determined in brain tissue by HPLC with simultaneous parallel–adjacent electrode system	Roston *et al.* (1982)
	Noradrenaline, adrenaline, dopamine, L-dopa	Determined in directly injected human urine (100 μl) using twin-electrode thin-layer cell. Upstream anode held at + 0.8 V (vs Ag/AgCl) and downstream cathode at + 0.2 V	Goto *et al.* (1982)
	Salsolinol (6,7-dihydroxy-1-methyl-1,2,3,4-tetrahydro-isoquinoline)	Determined in urine with graphite detector set at + 0.7 V (vs Ag/AgCl). Mean level determined in healthy adults: 1.89 $\pm$ 0.85 ng ml^{-1}	Dufay *et al.* (1991)

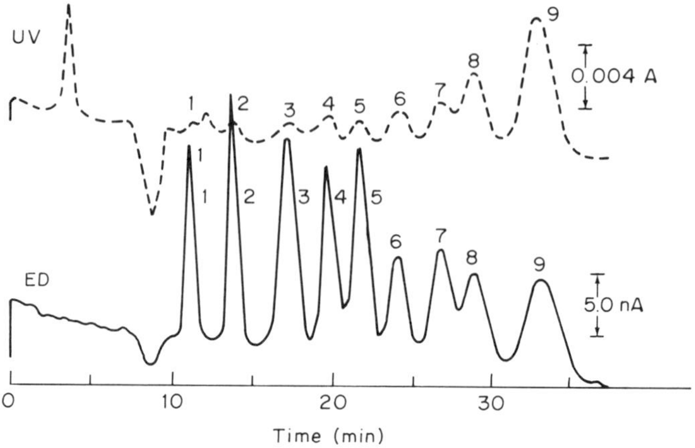

Figure 31. Simultaneous UV and electrochemical detection of phenols with Aminex resin. Column, 185 × 12.7 mm i.d.; eluent, 30% CH_3CN–0.025 M H_2SO_4 at 1.6 ml min^{-1}; UV wavelength, 213 nm; ED applied potential, + 1.20 V; sample size, 280 μl; Peak identities: (1) 117 ppb hydroquinone; (2) 166 ppb phenol; (3) 143 ppb *o*-chlor- and 186 ppb *p*-chloro-; (4) 285 ppb 2,6-dichloro; (5) 360 ppb 2,4-dichloro-; (6) 290 ppb 2,4,6-trichloro-; (7) 380 ppb 2,4,5-trichloro-; (8) 450 ppb 2,3,4,6-tetrachloro-; (9) 470 ppb pentachlorophenol. Reprinted with permission from *Analytical Chemistry*, **51** 1039. Copyright (1979) American Chemical Society

1985). The oxidation peak at + 0.57 V (vs Ag/AgCl), presumably corresponding to a hydroquinone oxidation, was used for quantitative purposes. A linear response was obtained for concentrations from 10^{-6} M to the LOD, 10^{-9} M, and it was possible to determine the drug in urine by direct injection of the sample with no pretreatment. The FIA–ED approach was found to be superior to a batch method in which the adsorption step was carried out in a stationary sample solution. This is illustrated in Fig. 33a, where the curve, obtained after a 3 min batch concentration, shows that the doxorubicin oxidation appears only as a barely discernible shoulder (peak 2) superimposed on the tail of the much larger peak ascribed to the oxidation of co-adsorbed uric acid. Although the latter compound adsorbs more weakly than doxorubicin, it exists at a substantially higher concentration in urine. The curve in Fig. 33b for the FIA–ED method shows that the adsorption of doxorubicin was virtually unaltered by the flow mode of sample presentation but that the weaker uric acid adsorption was much more drastically affected, resulting in resolution of peaks 1 and 2. Both adsorption approaches offer a distinct selectivity improvement over direct DPV, which provides no real discrimination against uric acid and other electroactive sample components, and hence is of little or no practical utility without extensive sample treatment.

Chastel *et al.* (1989, 1990) used lipid-modified carbon paste electrodes to accumulate the antitumour drug marcellomycin (**XXXVIII**) in the absence of an applied potential followed by oxidative voltammetric determination, utilising

Table 20. Determination of narciclasine in aqueous solutions and serum

Detection method	Sample	Limit of detection[a] (ng ml^{-1})	Linear dynamic range (μg ml^{-1})	Correlation coefficient	Precision (R.S.D.) (%)		
					0.1 μg ml^{-1}	0.5 μg ml^{-1}	1.0 μg ml^{-1}
Voltammetric	Aqueous	10	0.01–10	0.9936	6.5	3.2	1.4
(+1.8 V)	Serum	15	0.015–10	0.9991	7.2	4.5	1.9
Fluorimetric	Aqueous	25	0.015–5	0.9986	5.8	2.5	1.0
($\lambda_{exc.} = 360$ nm; $\lambda_{em.} = 480$ nm)	Serum	32	0.032–5	0.9984	6.1	3.2	1.5
UV photometric	Aqueous	15	0.015–12	0.9999	8.5	2.8	1.2
($\lambda = 254$ nm)	Serum	65	0.065–10	0.9944	10.4	3.5	1.6

[a]At a signal-to-noise ratio of 2.

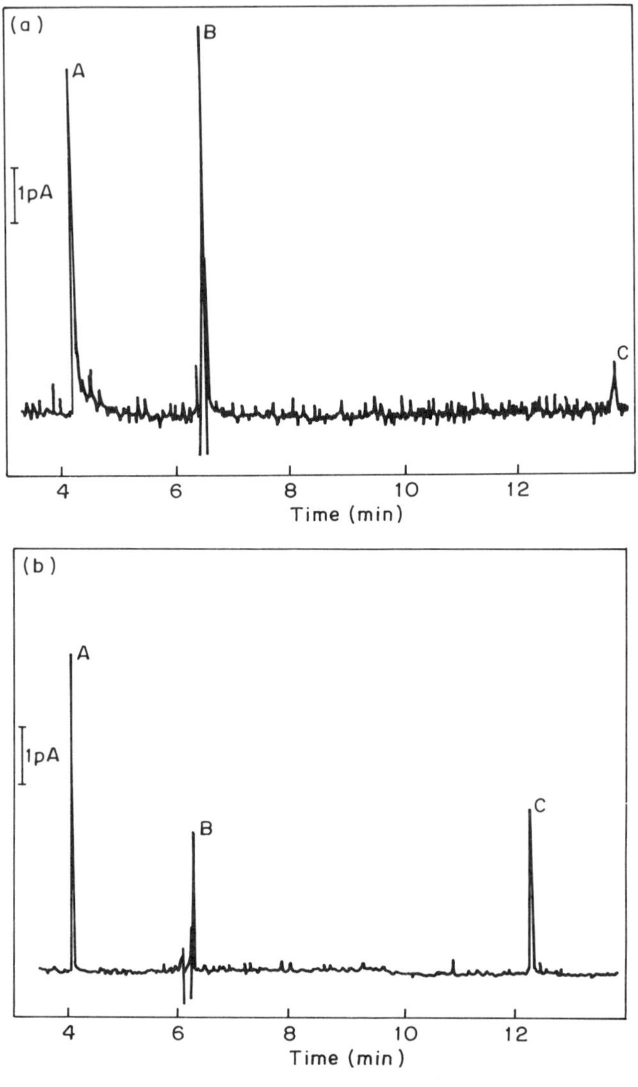

Figure 32. (a) Capillary electrophoretic separation of solutes in a sample of cytoplasm removed from inside the giant dopamine neuron of *Planorbis corneus*: electrophoresis capillary length, 66 cm; buffer, 0.025 M MES, pH 5.65; injection, 5 s at 10 kV through a 700 μm long, 10 μm o.d. microinjector; separation potential, 25 kV. Detection electrode, 0.7 V vs SSCE. The injection volume was 66 pl, based on electroosmotic flow. Peaks A, B and C have been identified as dopamine, a non-ionic species and dihydroxyphenylacetic acid based on their calculated electrophoretic mobilities. The electrophoretic mobility of peak A is 2.40 $cm^2\ V^{-1}\ s^{-1}$ and that of peak C is $-$ 2.28 $cm^2\ V^{-1}\ s^{-1}$. (b) Capillary electrophoretic separation of authentic standards obtained after the cellular experiment shown in (a). A = dopamine (105 amol), electrophoretic mobility is 2.40 $cm^2\ V^{-1}\ s^{-1}$; B = L-dihydroxyphenylalanine (69 amol) and C = dihydroxyphenylacetic acid (175 amol), electrophoretic mobility is $-$ 2.22 $cm^2\ V^{-1}\ s^{-1}$. The injection volume was 69 pl. Conditions were as in (a). Reproduced by permission of Elsevier Science Publishers from *J. Neurosci. Methods*, **34** (1990) 11

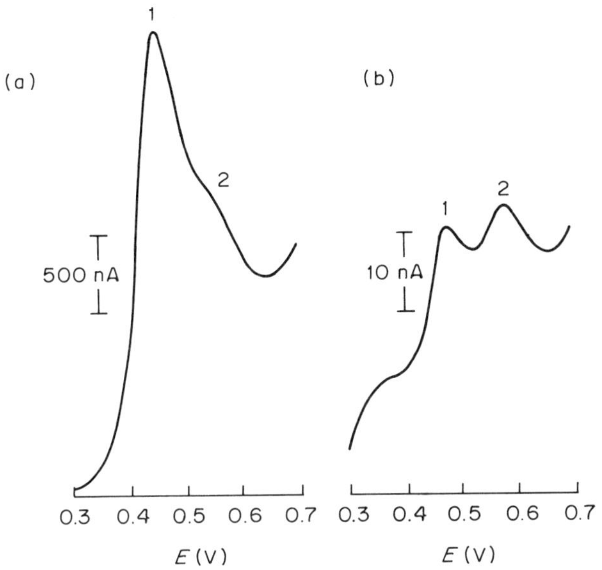

Figure 33. Differential-pulse voltammograms obtained by (a) batch preconcentration and (b) flow-injection preconcentration from a urine specimen with 5×10^{-7} M doxorubicin (6 μl injection at 0.1 ml min^{-1} flow rate). Reproduced by permission of Elsevier Science Publishers from *Anal. Chim. Acta*, **176** (1985) 105

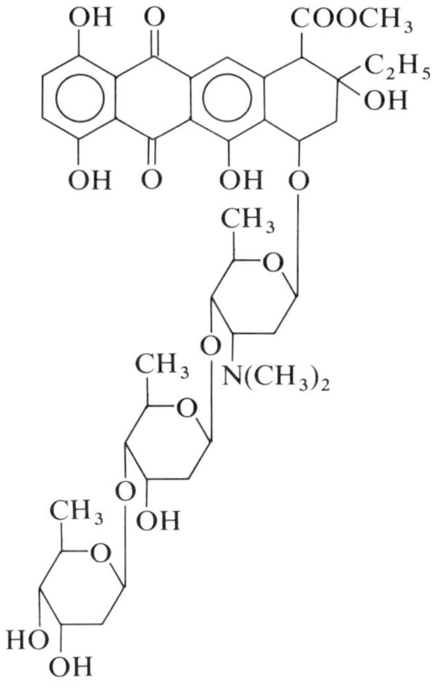

(XXXVIII)

the hydroquinone present in the overall molecular structure. The presence of the lipid can enhance the current up to eightfold and electrode response is linearly related to concentration in the range 1×10^{-8}–6×10^{-6} M.

An amperometric biosensor has been designed for the rapid determination of salicylate in as little as one drop of whole blood (Frew *et al.*, 1989). Salicylate hydroxylase converts salicylate to catechol in the presence of NADH and molecular oxygen. Catechol is monitored amperometrically by oxidation at +300 mV (vs Ag/AgCl) and the size of the oxidation current is related to the concentration of salicylate in the sample. All the reagents are incorporated into the working electrode of a disposable strip.

The anti-inflammatory drugs piroxicam (**XXXIX**) and tenoxicam can be accumulated at a lauric acid-modified carbon paste electrode and, together with the medium exchange procedure, this has resulted in an AdSV method for their determination in urine (El-Maali and Hassan, 1990).

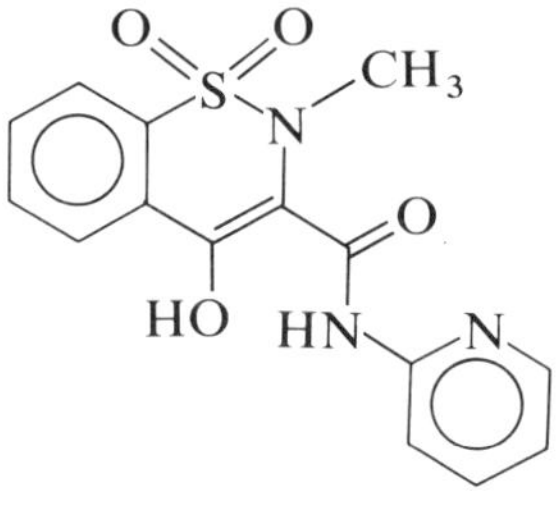

(**XXXIX**)

The elements Cu, Fe, U and V can be simultaneously determined by AdSV using an HMDE working electrode after chelation with catechol. Addition of 4×10^{-4} M catechol to sea water at pH 7.0, followed by 3 min accumulation of the metal chelates at − 0.05 V and their subsequent cathodic stripping using the differential-pulse mode, resulted in the appearance of four well separated peaks at − 0.23, − 0.40, − 0.56 and − 0.72 V for Cu(II), Fe(III), U(VI) and V(V), respectively. (Van den Berg and Huang (1984). Other metal ions such as Bi(III), Pb(II), In(III), Sb(V), Ni(II) and Zn(II) will also give stripping signals under these conditions and either their peak heights are significantly smaller for a given concentration of the metal ion or the selectivity is diminished by overlapping peaks, e.g. Sb(V) − 0.83, Ni(II) − 0.86 V. The optimum experimental conditions are different for each metal ion.

Ge(IV) has been adsorbed on an SMDE in the form of a cathechol chelate followed by differential-pulse voltammetric determination at − 0.61 V (vs Ag/AgCl) (Hasebe *et al.* 1990). Optimum solution conditions were 3.2 mM catechol, 0.7 M acetate buffer (pH 4.5) and 0.1 M sodium perchlorate. As(III) and Sn(II) interfered with the determination. Germanium was determined in Al–Ge alloys as shown in Table 21.

An AdSV technique was developed by Schleich and Henze (1990) based on the accumulation of Ge(IV)–diol complexes. Pyrogallol gave a limit of detection

Table 21. Determination of Ge in Al–Ge alloys by AdSV

Sample	Sample taken (mg)	Ge found (mg)	Concentration[b] (%)	Relative standard deviation (%)
Alloy I (12.4%)[a]	22.0	2.65 ± 0.12	12.1 ± 0.5	2.2
	25.3	3.07 ± 0.05	12.2 ± 0.3	
	20.7	2.60 ± 0.10	12.5 ± 0.5	
		Mean:	12.3 ($n = 9$)	
Alloy II (23%)[a]	10.3	2.59 ± 0.10	25.1 ± 0.9	1.8
	12.9	3.18 ± 0.13	24.7 ± 1.0	
	11.2	2.74 ± 0.12	24.4 ± 1.0	
		Mean:	24.7 ($n = 9$)	

[a]Certified values.
[b]95% confidence limit.

of 0.1 ng ml^{-1} with a relative deviation of 13%. The method was applied to the determination of Ge(IV) in ginseng and garlic.

3.3 CARBAMATES

Carbamates are polarographically reducible after derivatisation with nitrating and nitrosating agents (Rowe and Smyth, 1979). Alternatively, they can be oxidised at relatively high oxidation potentials as shown in Table 22.

Anderson and Chesney (1980) reported a reversed-phase HPLC method with thin-layer Kel-F-graphite electrochemical detection operated in the constant potential amperometric mode at + 1.10 V (vs Ag/AgCl). Calibration graphs were linear over at least three orders of magnitude with relative standard deviations of 1–2%. Detection limits of 2–7 ng ml^{-1} were obtained and the method compared favourably with GC with electron capture detection following hydrolysis to the corresponding phenols or amines and reaction with

Table 22. Oxidation potentials of carbamates

Carbamate	E_p (V vs Ag/AgCl)	Supporting electrolyte
Aminocarb (**XL**)	0.82	Phosphate buffer (pH 6)–30% CH_3CN–10% MeOH
Carbendazim	1.0	As above
Desmedipham (**XLI**)	0.94, 1.1	Phosphate buffer (pH 7.6)–10% MeOH
Carbaryl	1.2[a]	As above

[a]Ill-defined peak.

halogen-rich reagents. The electrochemical detector, in this case, was found to have lower LODs than UV detection at 190–210 nm or fluorescence detection in which dansyl derivatives were formed prior to injection or post-column derivatisation carried out with *o*-phthalaldehyde.

Anderson *et al.* (1985) further investigated the application of HPLC–ED to the determination of trace concentrations of carbamates in river water by the use of microarray electrochemical flow detectors operating at high applied potentials. These Kelgraf microarray detectors effectively discriminated between oxidation reactions limited by the rate of mass transport and reactions (including solvent oxidation) limited by the rate of electron transfer or other surface processes, and thus can afford improved LODs at high applied potentials. Sub-nanogram LODs were obtained for aminocarb (**XL**), carbendazim and desmedipham (**XLI**).

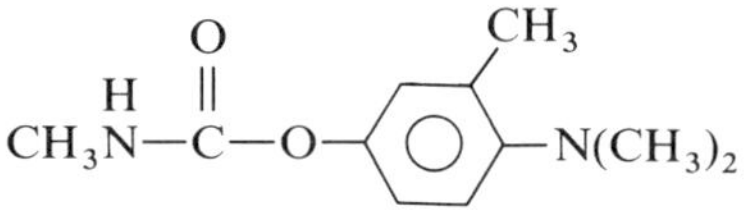

(**XL**)

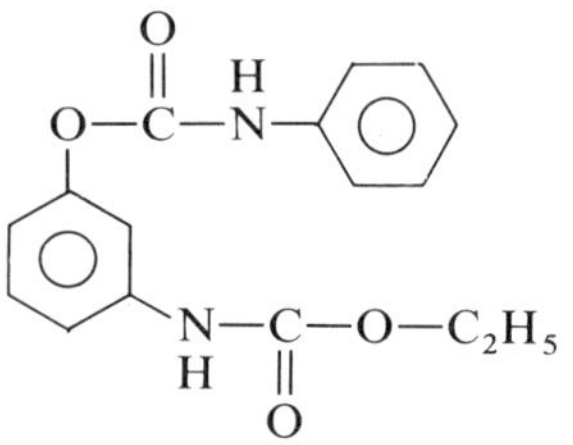

(**XLI**)

3.4 ORGANIC ACIDS, ALCOHOLS AND SUGARS

The polarographic behaviour of organic acids has been well documented in standard texts. In general, most carboxylic acids are reduced at relatively negative potentials [from − 1.50 to − 1.80 V (vs SCE)]. The main exceptions to this rule are those compounds which contain two carboxylic acid groups joined by an ethylene linkage, e.g. maleic and fumaric acids and the pyridinecarboxylic acids.

Santos and Baldwin (1986, 1987a) used a chemically modified electrode, cobalt phthalocyanine in carbon paste, as a detector in HPLC–ED for the determination of oxalic acid and α-keto acids, including pyruvic, phenylpyruvic, α-ketobutyric, α-ketoglutaric and α-ketoisocaproic acid. This chemically modified electrode catalysed the oxidation of oxalic acid and α-keto acids over the range 0.75 to 0.9 V (vs Ag/AgCl), which represents a roughly 500 mV decrease in overpotential compared with the use of an unmodified electrode. At + 0.75 V, the response for oxalic acid was linear from 1×10^{-7} to 2×10^{-5} M and gave a detection limit of 0.3 pmol. The detection limits for the α-keto acids varied from 0.15 to 1 nmol. Assay procedures using this detector in reversed-phase HPLC were developed for oxalic and pyruvic acid in plasma and urine samples. Results for samples from healthy volunteers and kidney-stone patients fell within the

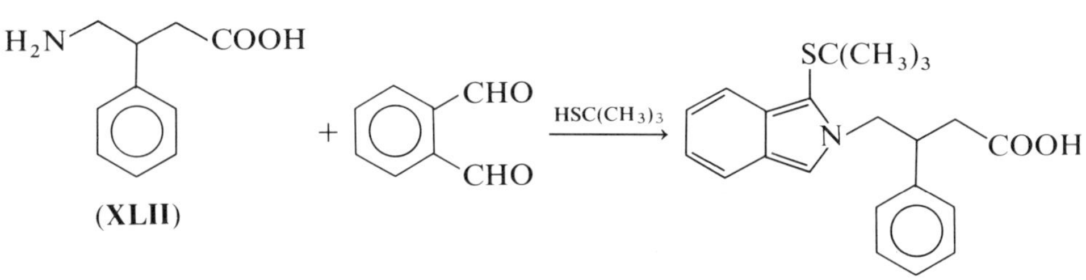

expected ranges for both analytes. The assay procedures were attractive because relatively low LODs were possible and the only sample treatment required was dilution and filtration.

There are relatively few instances in which polarographic methods have been applied to the determination of biologically important compounds containing the COOH functional group unless these compounds also contain another electroactive centre, which gives rise to waves/peaks more suited to trace analysis. In addition, derivatisation procedures can be employed, e.g. (i) 2,4-dichlorophenoxyacetic acid can be determined in blood and urine after nitration and TLC removal of interfering substances in the chloroform extract prior to application to toxicological situations, and (ii) baclofen (**XLII**) and α-baclofen can be determined in rat liver homogenate and human urine used for pharmacokinetic studies by solid-phase extraction, *o*-phthalaldehyde-*tert*-butyl thiol

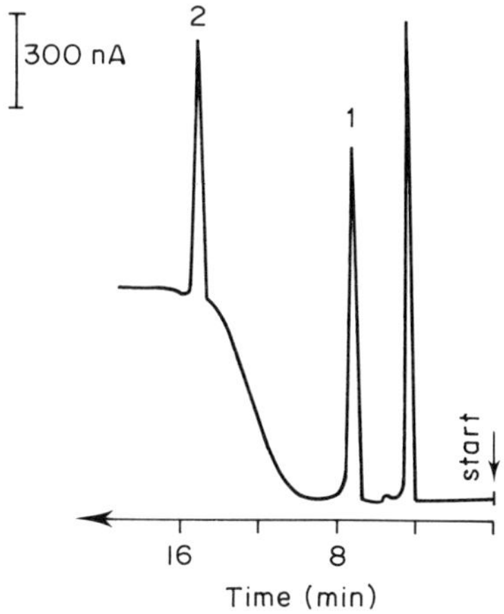

Figure 34. HPLC of an electrophoresis bath for plating metals with polymers containing 1-5% (v/v) isopropanol and butoxyethanol, diluted 1:25. Peaks: 1, isopropanol; 2, butoxyethanol. Amperometric detection (nickel at 480 mV vs Ag/AgCl 3 mol l^{-1} KCl. Column: LiChrosorb RP-18, 250 × 4.6 mm i.d., 25 °C. Eluent: acetonitrile–water (step gradient: 10 + 90 → 30 + 70 flow rate 0.7 ml min^{-1}). Post-column addition of NaOH (0.3 mol l^{-1}, flow rate 0.3 ml min^{-1}). Injection volume: 20 μl. Reproduced from *Fresenius J. Anal. Chem.*, 1991 **339**, 55 by permission of Springer-Verlag, Heidelberg

Table 23. Application of chemically modified electrodes (CMEs) to sugar determinations

CME	Analyte	Summary
Reticulated vitreous carbon with covalently attached glucose oxidase (Wieck *et al.*, 1984)	Glucose	Glucose oxidase was immobilised on the surface of a flow-through reticulated vitreous carbon electrode by reaction with a water-soluble carbodiimide. In FIA, injections of glucose samples resulted in the enzymatic production of H_2O_2 at the CME surface. By holding the CME at +0.9 V (vs SCE) and monitoring the anodic current due to oxidation of the H_2O_2 so produced, glucose could be quantified over the 10^{-4}–10^{-1} M range. The CME response was linear for glucose concentrations from 2.5 to 10 mM, with a relative standard deviation of less than 2%
Meldola Blue adsorbed on graphite (Appelqvist *et al.*, 1985),	Glucose	NADH was oxidised electrocatalytically at the CME at 0 V (vs SCE) which was at least 300 mV less than the potential required at a bare graphite electrode. Glucose was quantified by inserting a packed-bed glucose dehydrogenase reactor ahead of the CME in a flow stream containing NAD^+ and measuring the NADH produced on sample injection. The detection limit was 0.25 mM, and the response was linear up to 1 mM. Because of the low potential possible with the CME, potential interferents, such as ascorbic acid, uric acid and catechols, posed no problem
Anthracene–ferrocene ester and glucose oxidase adsorbed on graphite (Jonsson *et al.*, 1989)	Glucose	The modifier had a dual function: the anthracene group was strongly adsorbed on the graphite substrate to immobilise the modifier, while the ferrocene moiety was able to mediate the electron transfer with glucose oxidase. Subsequently, glucose oxidase itself was adsorbed on top of the anthracene–ferrocene derivative. In FIA, the CME exhibited an anodic response for glucose at the redox potential of the ferrocene compound. This response was suitable for glucose detection from 50 μM up to 1.5 mM. Because the enzyme itself was electroactive at this CME surface, detection could be performed without the addition of an oxidizing agent or cofactor to the mobile phase

Bis(benzophenoxazine) derivative adsorbed on graphite (Marko-Varga, 1989)	Mono- and oligosaccharides	A CME containing an immobilised phenoxazine gave electrocatalytic oxidation of NADH at 0 V (vs Ag/AgCl). Carbohydrate samples were first separated on a Ca-loaded ion-exchange column and then passed through two enzyme reactors containing amyloglucosidase and glucose dehydrogenase–mutarotase. NADH formed in the latter enzymatic process was detected at the CME located downstream. Sugars up to maltohexose were quantified at the 100–300 ng level in samples including soft drinks, beer and fermentation broths
Glucose oxidase and platinum particles co-deposited on carbon fibres (Wang *et al.*, 1989)	Glucose	This CME was constructed by electrochemically reducing hexachloroplatinate from a solution also containing glucose oxidase. In the presence of glucose, the enzymatic reaction at the CME resulted in the production of H_2O_2, which was oxidized with low overpotential at the platinum metal. Glucose concentrations as low as 1.3×10^{-5} M were accessible with a relative standard deviation of 3 %, and continued monitoring with the same surface was possible for at least 20 days.
CoPC in carbon paste (Santos and Baldwin, 1988)	Carbohydrates and other polyhydroxy compounds	Electrocatalysis of carbohydrates was extended to include oligosaccharides, deoxy sugars and simple polyalcohols. Detection (with pulsed applied potential) was performed in strongly basic solution following anion-exchange chromatography. Detection limits were in the 10–50 pmol range. Sample matrices included soft drinks, beer and urine
CoPC in carbon paste (Santos and Baldwin, 1987b)	Carbohydrates	The electro-oxidation of sugars occurred at low potentials in strongly basic solution. The carbohydrates examined included mono- and disaccharides, pyranose and furanose rings and reducing and non-reducing sugars. Use of this phenomenon in HPLC–ED required that the applied potential was pulsed between +0.4 V (for detection) and −0.3 V (for electrode re-activation) (vs Ag/AgCl). Coupled with anion-exchange chromatography in 0.15 M NaOH, the CME provided detection limits of 100–500 pmol. The stability of the pulsed detection was such that the relative standard deviation for 40 injections over 5 h was only 2 %
CoPC in carbon paste (Tolbert and Baldwin, 1989)	Carbohydrates, alditols and acidic sugars	CoPC CMEs catalysed the oxidation of alditols and aldonic, uronic and aldaric acids similarly to that of simple carbohydrates. For HPLC–ED, detection was at +0.42 V (vs Ag/AgCl), but the potential was continually pulsed to −0.3 V to regenerate the active form of the CoPC catalyst. For anion-exchange chromatography in strongly alkaline solution, detection limits ranged from 10 pmol for the alditols up to the nmol level for some of the aldaric acids

derivatisation as shown and HPLC–ED at a glassy carbon electrode held at + 0.60 V (vs Ag/AgCl). The method was shown to be quantitative in the range 10 ng ml^{-1}–10 μg ml^{-1} (Wall and Baker, 1989).

Torrance and Gatford (1987) developed a method for the direct determination of soluble chromium at concentrations < 2 μg l^{-1} in simulated PWR coolant by DPAdSV. The method is based on measurement of the current associated with reduction of a Cr(III)–DTPA chelate (DTPA = diethylenetriaminepentaacetic acid). No interference was observed from other transition metal ions expected to be present in PWR coolant. This methodological approach could also serve for the determination of chelating carboxylic acids.

Alcohols are not directly amenable to polarographic analysis unless they contain another functional group that is electroactive, e.g. 2,3-dimercaptopropan-1-ol (British Anti-Lewisite, BAL). BAL can be determined in blood serum by a method based on monitoring the oxidation of the SH groups in the molecule. Other polarographic methods have been described for some biologically important alcohols (Van Beeumen and Deley, 1971).

Stitz and Buchberger (1991) found simple alcohols, polyols and hydroxycarbonic acids to be detectable by amperometric detection at nickel electrodes subsequent to HPLC. For quality control purposes, electrophoresis baths used for plating metals with polymers had to be analysed for isopropanol and butoxyethanol in the range 1–5% (v/v). These complex samples with partly unknown composition were of high viscosity and contained pigments and polymerising agents which are troublesome in GC determinations. HPLC with a C_{18} reversed-phase column and detection at a nickel electrode proved a satisfactory solution to the problem (Fig. 34). Selected sugar and sugar degradation products were also measured by this technique.

Sugars give rise to kinetic waves/peaks at relatively negative potentials from − 1.5 to − 1.7 V but the resulting currents are generally too small for trace analytical purposes. They can, however, be conveniently determined using chemically modified electrodes, e.g. the ferrocene-mediated enzyme electrode for amperometric determination of glucose in diabetic blood samples (1–30 mM) (Cass *et al.*, 1984). Other examples of the application of chemically modified electrodes to sugar determination are given in Table 23. In addition, Andrews and King (1990) used pulsed amperometric detection (PAD) for carbohydrates at gold electrodes following HPLC and Martens and Frankenberger (1991) used anion-exchange HPLC–PAD for the determination of aminosaccharides, in particular. HPLC–PAD required less sample preparation, was more precise and gave a nearly two orders of magnitude lower LOD than HPLC with a refractive index detector. Martens and Frankenberger (1991) found that the concentration of total aminosaccharides directly extracted from selected animal wastes and plant residues ranged from 0.46 g kg^{-1} (poultry manure) through 0.64 g kg^{-1} (sewage sludge) and 13.5 g kg^{-1} (alfalfa) to 17.2 g kg^{-1} for straw (table 24). The results also indicated that about 3.7% of the total nitrogen in alfalfa and 20% of that in straw was present as aminosaccharides. Zadeii *et al.* (1991) recently

Table 24. Composition and amount ($mg\ kg^{-1}$) of aminosaccharides extracted from chitin, organic amendments, organic-amended soil and microbial polymers

Aminosaccharide	Chitin	Poultry manure (PM)	PM-treated soil	Sewage sludge (SS)	SS-treated soil	Straw (ST)	ST-treated soil	Alfalfa (AL)	AL-treated soil	Check soil	*Chromo bacterium violaceum*	*Hansenula holstii*
Galactosamine	ND[a]	49	4	188	5	1.41×10^3	5	1.83×10^3	13	1	2.06×10^4	ND*
Mannosamine	ND	39	3	90	3	230	3	685	2	< 1	7.02×10^3	ND
Glucosamine	7.15×10^4	38	4	115	5	905	5	1.21×10^3	34	2	8.63×10^3	ND
N,N′-Diacetylchitobiose	ND	187	6	84	3	1.31×10^4	4	6.04×10^3	5	1	1.27×10^4	1.09×10^3
Acetyl-galactosamine	ND	18	< 1	35	1	149	13	245	1	1	2.38×10^3	ND
Acetyl-mannosamine	ND	67	2	73	3	390	3	1.57×10^3	0	1	1.96×10^4	ND
Acetyl-glucosamine	1.21×10^5	60	3	59	2	968	3	1.95×10^3	1	< 1	1.23×10^4	ND
Total	1.93×10^5	458	23	644	22	1.72×10^4	36	1.35×10^4	56	7	8.33×10^4	1.09×10^3

[a]ND, not detected.

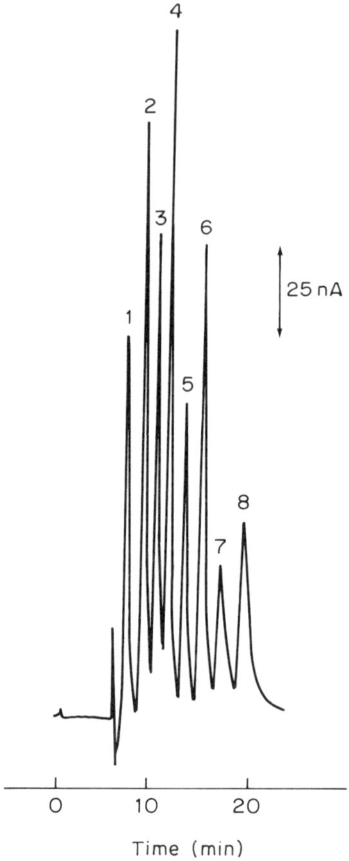

Figure 35. Chromatograms of mixtures of carbohydrates containing (1) inositol, (2) sorbitol, (3) rhamnose, (4) arabinose, (5) glucose, (6) ribose, (7) lactose and (8) sucrose. Carbohydrate concentrations: 3×10^{-4} M (1, 2, 4), 2×10^{-3} M (3, 5, 7, 8) and 5×10^{-3} M (6). Injection volume, 20 μl. Flow rate, 0.3 ml min^{-1}. Applied potential, 0.48 V. Mobile phase, 0.15 M NaOH. Reprinted with permission from *Analytical Chemistry*, **63**, 649. Copyright (1991) American Chemical Society

used copper/copper oxide particles dispersed in a Nafion film that is cast on the surface of a glassy carbon electrode as a detector for the HPLC separation of carbohydrates. A separation of eight carbohydrates is shown in Fig. 35.

4

Voltammetric Analysis of Selected Hydrocarbons Containing Carbon–Carbon Multiple Bonds

The carbon–carbon double bond is generally very difficult to reduce, particularly when it is contained within an aromatic nucleus. For example, polynuclear aromatic hydrocarbons are reduced between − 2.00 and − 2.50 V (vs SCE) in 0.175 M Bu_4NI–75% dioxane. As a result, there have been relatively few examples in which such reductive polarographic methods have been applied to the determination of relevant compounds. On the other hand, much information has appeared in the literature on the oxidation of vitamin C or ascorbic acid, which involves conversion of the —C(OH)=C(OH)— functional group to —C(=O)—C(=O)—. Davidson (1979) reviewed the application of this oxidation process to formulation analysis.

Lau *et al.* (1989) recently simultaneously determined ascorbic acid, caffeine and paracetamol in formulations by DPV at a glassy carbon electrode. Reversed-phase HPLC can be used to separate ascorbic acid from interferences in biological fluids prior to ED. Iriyama *et al.* (1985) carried out a preliminary protein precipitation on body fluids with metaphosphoric acid–EDTA. HPLC was then applied with a mobile phase of 0.1 M potassium phosphate–citric acid buffer (pH 4.6) containing 1 mM EDTA with ED at a glassy carbon electrode held at + 0.80 V (vs Ag/AgCl). Uric acid could be measured simultaneously. The concentrations of ascorbic acid in human serum, urine and cerebrospinal fluid were 12.8, 36.1 and 9.01 $\mu g\ ml^{-1}$, respectively. Ascorbic acid can also be measured in brain tissue by this same approach of HPLC–ED (Thrivikraman *et al.*, 1974).

Cephalosporin antibiotics such as cephalothin (**XLIII**) contain a substituted C=C group and polarographic reduction is believed to result in cleavage of the CH_3COO group (Hall *et al.*, 1977) with an overall consumption of $2e^-$. Peled *et al.* (1987) utilised the adsorption of **XLIII** on mercury electrodes to produce an

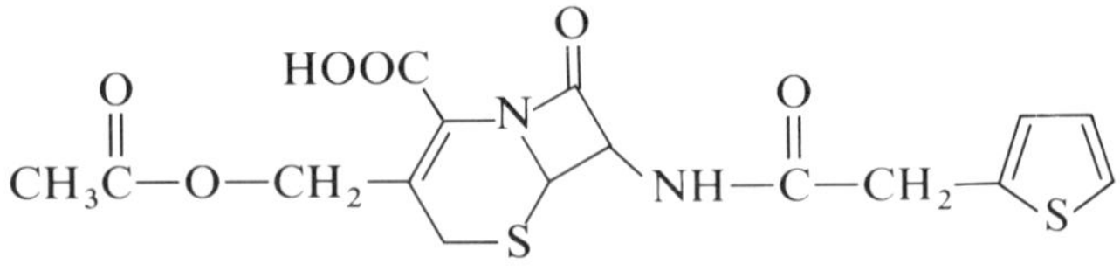

(XLIII)

AdSV method for its determination in 0.075 M $HClO_4$ down to 10^{-8} M concentrations. Similar LODs were reported by Munoz *et al.* (1989) for cefazolin and cefmetazole. Lower concentrations are not attainable owing to hydrogen evolution catalysed by adsorbed cephalothin, which masks the cephalothin reduction signal. Such a method is not applicable to the direct analysis of biological fluids owing to extensive interference from naturally occurring constituents of the particular biological fluid. This can be circumvented by introducing a separation stage prior to electrochemical analysis. Altinoz *et al.* (1990) extracted ceftriazone from blood plasma with the non-ionic resin Amberlite XAD-2 and the antibiotic was eluted finally from the column with methanol. The methanol was evaporated and the residue reconstituted in Britton–Robinson BR buffer (pH 3). AdSV in the differential-pulse mode was then used to determine ceftriazone concentrations in maternal and umbilical cord blood in caesarian cases. The amount of the drug transmitted to the fetus by administration to the mother before the caesarian operation was found to be in the range 0.067–0.17%. HPLC has also been used as the separation process. For example, Selavka *et al.* (1986) used HPLC–*hν*–ED to attain an LOD of 40 ppb for the cephalosporin cefoperazone. They claimed that the improved selectivity of this technique is due to three factors: (a) HPLC capacity factors or retention times, (b) photolytic lamp-on and lamp-off qualitative responses and (c) dual-electrode response ratios under both lamp-on and lamp-off conditions.

Vitamin A, with its conjugated carbon–carbon double bond structure, can be conveniently determined by low-pressure LC–ED using a wall-jet cell with a glassy carbon electrode operated in the oxidative mode (Hart and Jordan, 1989). Vitamin A has been so determined in a multivitamin preparation using a mobile phase of 65% methanol–0.075 M acetate buffer (pH 5.0). The relative standard deviation was found to be 3.8% ($n = 10$) and the LOD was 300 pg injected.

Compounds containing the —C(=O)—C=C— functional group are reduced polarographically at more positive potentials than a single carbon–carbon double bond. 3-Keto steroids with unsaturation between C-4 and C-5, e.g. cortisone and testosterone, give well defined DPP peaks for the C=C reduction at − 1.77 and − 1.84 V (vs SCE), respectively, in 0.1 M Et_4NClO_4–50% methanol. AdSV was used by Wang *et al.* (1985a) to obtain LODs of 1.6×10^{-10} M for testosterone, 2×10^{-10} M for progesterone and 3.3×10^{-10} M for methyltestosterone with 15 min preconcentration at the SMDE. Norethisterone (17α-ethynyl-17β-hydroxyoestr-4-en-3-one) is used in oral contraceptives, and can be rapidly assayed in tablets by DPP without the need for prior extraction

(Opheim, 1977). Multisteroidal formulations are, however, best determined by HPLC. For example, Capitano (1979) used reversed-phase HPLC with UV detection at 280 nm to separate and determine equilin, equilenin, oestrone and oestradiol in a formulation, whereas Roos (1978) separated the dansyl derivatives of oestrone, oestradiol valerate, ethinyloestradiol, α-oestradiol and oestradiol, followed by fluorimetric determination in 15 min. The cardiovascular drugs digoxin and digitoxin also contain the basic steroidal nucleus and polarographic reduction occurs at the C=C bond in the five-membered ring containing a conjugated carbonyl group. The LOD of these compounds by DPP has been reported as 2.5×10^{-6} M (Kadish and Spiehler, 1975), which can be improved by the application of AdSV down to 10^{-9}–10^{-10} M following 15 min accumulation at the SMDE (Wang *et al.*, 1985c). Digoxin concentrations in plasma in the low and narrow therapeutic range (0.8–2.0 ng ml^{-1}) have been determined by Wehmeyer *et al.* (1986) using a heterogeneous enzyme immunoassay based on the ELISA technique. Alkaline phosphatase, which catalyses the hydrolysis of the phenyl phosphate ester to phenol and phosphate, is used as the enzyme label on the drug and it is this enzyme-labelled drug which competes with the free drug for binding with the immobilised antibody. The phenol is easily detected by HPLC–ED with a carbon paste electrode held at + 0.87 V (vs Ag/AgCl) and the calibration graph of this current vs digoxin concentration is typically non-linear. The LOD is as low as 50 pg ml^{-1} and results of analyses on patients' samples were found to be in good agreement with those of an accepted radioimmunoassay procedure. Visconti *et al.* (1984) investigated the DPP of trichothecene mycotoxins, which again contain the α,β-unsaturated carbonyl

Table 25. DPP and GC determination of deoxynivalenol, nivalenol and fusarenone X in extracts of maize samples, artificially or naturally infected

	Trichothecene (μg g^{-1})					
	Deoxynivalenol		Nivalenol		Fusarenone X	
Sample	DPP	GC	DPP	GC	DPP	GC
Culture of:						
F. graminearum						
(ITM126)	1.18	1.20	—	—	—	—
F. equiseti						
(ITM5)	ND[a]	ND	—	—	—	—
F. tricinctum						
(ITM16)	ND	ND	9.80	10.50	2.0	2.10
Fusarium-infected maize from:						
Austria	—	—	0.50	0.70	0.44	0.66
N. Italy	20.00	18.00	—	—	—	—

[a]ND = not detected

functional group. They determined nivalenol, deoxynivalenol and fusarenone X in infected maize using a suitable liquid–liquid extraction followed by chromatographic clean-up prior to DPP. The results were compared with those obtained by a GC method (Table 25).

The carbon–carbon triple bond is not directly reducible at the DME. With the artificial steroid lynestrenol, heating with a 60% H_2SO_4–methanol mixture converted the $-C{\equiv}CH$ group into a polarographically active moiety (Van Bennekom *et al.*, 1975).

5
Voltammetric Analysis of Halogen-containing Molecules

Aromatic halogen-containing organic compounds can be reduced according to

$$RX + 2e^- + H^+ \rightarrow RH + X^-$$

with the reduction potentials being markedly affected by the nature and number of substituents in the aromatic ring. Although organochlorine compounds are best determined using GC with mass spectrometric, electron-capture or coulometric detection, there are several examples in the literature in which voltammetric methods have found particular application, e.g. Kemula and Kreminska (1960) combined column chromatography and polarography to separate quantitatively the *p,p'* and *o,p'*-isomers of DDT; Davidek and Janicek (1961) nitrated DDT in a biological matrix to give the tetranitro derivative, resulting in an analytical method with an LOD 0.4 μg ml^{-1}; Supin and Budnikov (1973) used the adsorption properties of γ-benzene hexachloride (γ-BHC, lindane) at a slow DME to determine it at levels down to 60 ppb in pure solution; Colas *et al.* (1964) used silica gel in order to separate a mixture of the two isomers of endosulfan, followed by polarography in a water–acetone mixture (50:50) containing LiCl. This compound was detected polarographically in the stomach contents of poisoned chickens in this investigation. Farwell *et al.* (1975) reported the use of interrupted-sweep voltammetric analysis for the identification of polychlorinated insecticides and other polychlorinated aromatics (Table 26, Fig. 36) The apparatus consisted of a three-electrode, potentiostatically controlled circuit with a logic-controlled interrupted linear-sweep mechanism. By using 0.1 M Et_4NBr–DMSO as the supporting electrolyte, voltammograms for many polychlorinated aromatic compounds have been obtained. This method suffers from the fact that at least 9 μg of relatively pure compound are required to obtain positive identification. It is therefore of little use in the determination of pesticide residues in the environment, but could find more application in the identification of isomeric products found during the synthesis of polychlorinated agrochemicals.

Table 26. Reduction potentials of some organochlorine insecticides using the method of Farwell *et al.* 1975)

Compound	Interrupt potential, E_{2d}[a] (V vs SCE)			
Heptachlor	−1.656,	−2.087,	−2.319	
Dieldrin	−1.675,	−2.038,	−2.233,	−2.450
DDE	−1.814,	−1.935,	−2.177,	−2.506
Chlordane	−1.753,	−2.106,	−2.333	

[a]The analogue second derivative of the current was used to control the voltages at which the scans were interrupted.

Chatten and Amankwa (1985) investigated the reduction mechanism of anti-herpes compounds, (*E*)-5-(2-bromovinyl)-2′-deoxyuridine and 5-iodo-2′-deoxyuridine and found that the former compound is reduced at − 1.80 V (vs SCE) in a $4e^-$ process involving loss of the bromine atom and reduction of the vinyl group. The latter compound is reduced in pH 7 buffer at − 0.97 V (vs SCE) owing to loss of the iodine atom.

Van Kerchove *et al.* (1982) compared the accuracies obtained in DPP of two different ways of establishing the calibration graphs for tablets containing a mixture of hydrochlorothiazide (**XLIV**) (which reduces by loss of the 6-Cl atom) and dihydralazine sulphate, which are reduced at potentials differing by 200 mV.

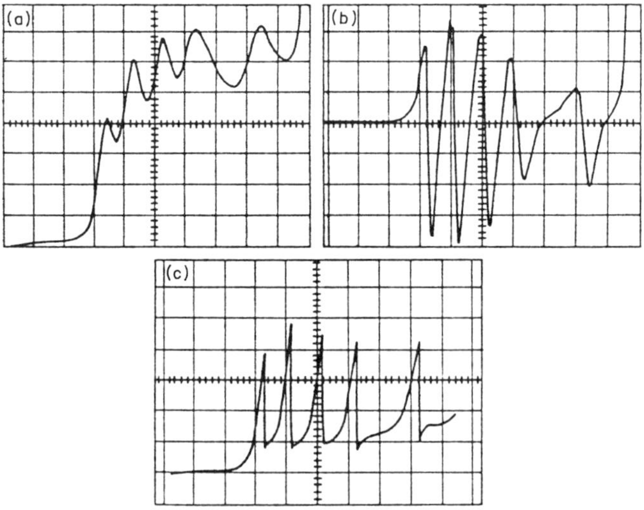

Figure 36. Reduction voltammograms of 1,2,3,4-tetrachloronaphthalene (voltage range − 0.7 to − 2.6 V (vs SCE) (a) Normal scan at 0.5 V per division and 1000 μA full-scale, (b) second-derivative scan at 1.0 V per division and 1000 μA full-scale; (c) interrupted-sweep fingerprint at 0.5 V per division and 1000 μA full-scale. Reprinted with permission from *Analytical Chemistry*, **47**, 895. Copyright (1975) American Chemical Society

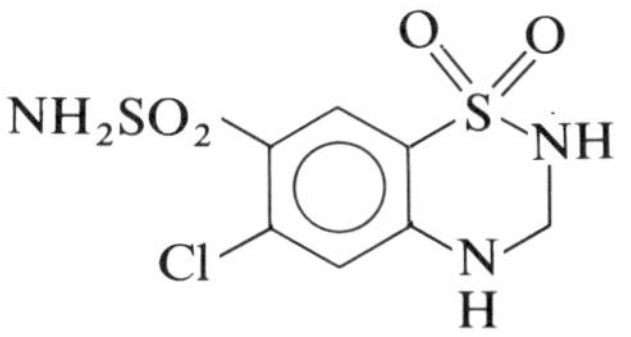

(XLIV)

The recommended procedure is to transfer a single tablet (equivalent to 10 mg of each electroactive species) into a 50 ml volumetric flask, grind it thoroughly and add Michaelis borate buffer (pH 9.7) followed by shaking for 30 min in the dark to prevent degradation of dihydralazine. The solution is filtered and 6 ml of the filtrate are transferred into a 20 ml volumetric flask. This is diluted to the mark with buffer and mixed prior to deaeration and DPP over the potential range – 0.6 to – 2.0 V (vs SCE). The diffusion currents at both peaks are measured and the amounts of the active ingredients determined by a calibration procedure with a zero covariance and also by a calibration procedure in which

Table 27. Effect of the addition of (a) 500 μg and (b) 1 mg of excipients on the stripping currents and potentials of 5 ml of solution of compound **XLVII** at a concentration of 3×10^{-5} M in 0.05 M KNO_3 —HNO_3 (9 μg ml^{-1}). Reprinted with permission from *Analytical Chemistry*, **51**, 2127. Copyright (1979) American Chemical Society

	Before excipient addition		After excipient addition		
	Peak current,		Peak current (μA)		
Excipient	μA	E_p(V)[a]	a	b	E_p (V)[a]
Lactose USP	2.23	+0.20	2.10	2.00	+0.20
Talc USP	2.15	+0.20	2.18	2.18	+0.20
Povidone USP	2.08	+0.20	1.50	1.65	+0.19
Maize starch BP	2.78	+0.20	2.73	2.58	+0.20
Avicel	2.28	+0.20	2.33	2.05	+0.20
Solka Floc	3.28	+0.20	2.88	3.75	+0.21
Amijel	2.75	+0.20	3.00	2.68	+0.17
Celacol 20 M	3.00	+0.20	3.28	4.00	+0.16
Amberlite IRP 88	2.45	+0.20	2.75	1.35	+0.20
Magnesium stearate USP	2.35	+0.20	2.58	2.55	+0.20
Dibasic calcium phosphate	2.10	+0.20	2.10	2.18	+0.20
Sodium starch glycolate	2.55	+0.20	⪢5.	⪢5.0	+0.20
Sodium lauryl sulphate BP	3.45	+0.20	Complete signal removal		
Soft gelatin mix 1	1.80	+0.20	1.43	1.43	+0.15
Soft gelatin mix 2	2.75	+0.20	2.55	2.80	+0.14
Soft gelatin mix 3	1.73	+0.20	1.98	2.00	+0.15
Hard gelatin mix	2.10	+0.20	1.93	1.95	+0.15

[a]Vs SCE.

Table 28. Comparison of stripping voltammetric and UV spectrophotometric results for dissolution of compounds **XLVII** and **XLVIII** from tablets. Reprinted with permission from *Analytical Chemistry*, **51**, 2127. Copyright (1979) American Chemical Society

	Cathodic stripping voltammetry			UV spectrophotometry		
	Concentration of drug in original tablet					
Compound	Measurement method a[a]	Measurement method b[a]	Concentration level in voltammetric cell (M)	Molar absorptivity water ($m^2\,l\,mol^{-1}$)	Concentration of drug in original tablet	Concentration level measured, (M)
XLVII	Calculated 19.7 mg (mean of 9 results) R.S.D. 3.1% Actual content 19.6 mg (100.5% recovery)	Calculated 19.7 mg (mean of 9 results) R.S.D. 3.5% Actual content 19.6 mg (100.5% recovery)	2×10^{-5}	2174 (276 nm)	Calculated 20.6 mg (mean of 10 results) R.S.D. 1.7% Actual content 20.7 mg (99.5% recovery)	2.5×10^{-4}
XLVIII	Calculated 16.5 mg (mean of 7 results) R.S.D. 2.9% Actual content 16.4 mg (100.5% recovery)	Calculated 16.0 mg (mean of 7 results) R.S.D. 2.5% Actual content 16.4 mg (97.5% recovery)	5×10^{-6}	2073 (280 nm)	Calculated 16.6 mg (mean of 10 results) R.S.D. 1.0% Actual content 16.6 mg (100.0% recovery)	2.5×10^{-4}

[a]Method a, measuring peak height from starting current to peak top; method b, measuring peak area by triangulation.

the standards contain mixtures with a constant concentration ratio of hydrochlorothiazide to dihydralazine of 1.1. The first calibration procedure in which the covariance C is made equal to zero in order to minimise experimental errors, i.e.

$$C = \frac{\sum AB - (\sum A \times \sum B)/n}{n - 1} = 0$$

where A and B are the concentrations of the two compounds in n solutions, was chosen since the relative standard deviations for both compounds were significantly improved, e.g. 0.88 % as opposed to 3.44 % for hydrochlorothiazide. When used for quality control on ten tablets, the mean value was 10.82 mg per tablet with a relative standard deviation of 4.78 %, which met the requirements of the USP (XX Revision) for content uniformity carried out on ten tablets, i.e. 85.0–115.0 %.

This polarographic method for hydrochlorothiazide (**XLIV**) should be critically compared with the HPLC–UV stability indicating assay of Daniels and Vanderwielen (1981), which can simultaneously determine likely degradation and process impurities such as chlorothiazide (**XLV**) and 4-amino-6-chlorobenzene-1,3-disulphonamide (**XLVI**).

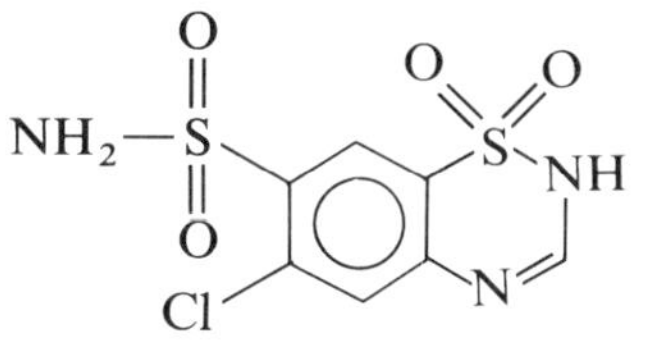

(**XLV**)

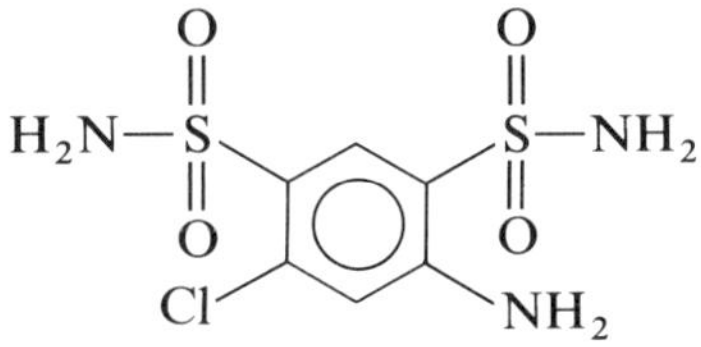

(**XLVI**)

A seldom exploited, although sensitive, analytical technique for halide measurement is cathodic stripping voltammetry (CSV). Work to date has been limited mainly to inorganic analysis. A method was developed by Davidson and Smyth (1979) to determine therapeutically active organic hydrochlorides and hydrobromides in drug dissolution studies by CSV. Two model compounds are used to illustrate the application of CSV, viz. (−)-*cis*-2-(α-dimethylamino-*m*-hydroxybenzyl)cyclohexanol hydrochloride (**XLVII**) and (−)-13β-amino-5,6,7,8,9,10,11,12-octahydro-5α-methyl-5,11-methanobenzocyclodecen-3-ol hydrobromide (**XLVIII**). Results on tablets were accurate at concentrations of

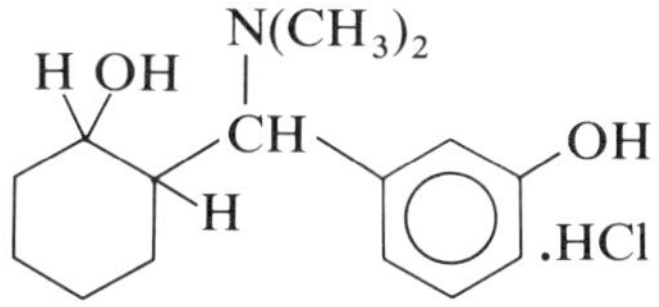

(**XLVII**)

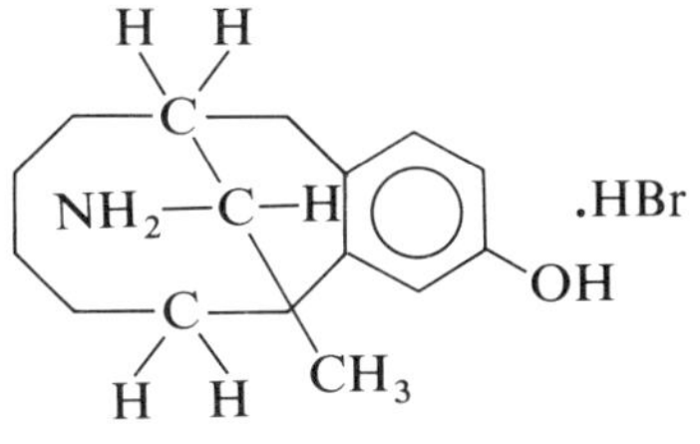

(**XLVIII**)

1×10^{-5} and 5×10^{-6} M for hydrochlorides and hydrobromides with relative standard deviations of 3.1 and 2.9%, respectively. Problems associated with halide monolayer formation on hanging mercury drop electrodes were considered, and by presaturating the supporting electrolyte with mercury(I) halide limits of detection of 1×10^{-6} and 2×10^{-7} M for chloride and bromide, respectively, were obtained in 0.05 M nitric acid–potassium nitrate. The effects of 17 common pharmaceutical excipients were studied (Table 27) and the method was shown to have an order of magnitude lower LOD than UV spectrophotometry for drugs with molar absorptivities of 2000 $m^2\ lmol^{-1}$. Table 28 compares the results of CSV and UV spectrophotometry for samples withdrawn from a tablet dissolution test using equipment described in the US Pharmacopeia (1975) and a dissolution medium of water.

6

Voltammetric Analysis of Organophosphorus Molecules

With the large-scale development of organophosphorus insecticides in the early 1950s, methods of analysis were based either on colorimetric techniques or on the determination of total phosphorus. These methods are inherently non-specific, and were superseded by the development of GC procedures employing flame photometric or thermionic detection. For example, Le Bel *et al.* (1979) determined organophophorus compounds in water in the concentration range 10–100 ng l^{-1} and Hild *et al.* (1978) determined these compounds in food down to a detection limit of 0.5 mg kg^{-1}. 'Cold' methods of analysis such as polarography/voltammetry are particularly useful when the organophosphorus compound is unstable under the particular GC conditions. HPLC is also of considerable value in this context. The recently published method of Di Corcia and Marchetti (1991) using graphitised carbon black cartridge extraction prior to HPLC with UV detection gave LODs of 0.07 $\mu g\ l^{-1}$ for paraoxon, fenitrothion, 0.06 $\mu g\ l^{-1}$ for ethyl parathion and 0.01 $\mu g\ l^{-1}$ for ethyl azinphos.

Nangniot (1964) studied the polarographic behaviour of a wide range of organophosphorus pesticides. Although they cannot be reduced at the DME, thiophosphates can produce sharp adsorption peaks using a fast-scan linear-sweep voltammetric technique. The peaks obtained for ten important organophosphorus compounds using a scan rate of 250 mV s^{-1} are shown in Fig. 37. Since thiophosphates give rise to anodic waves/peaks at the DME owing to the formation of partially insoluble mercury compounds, CSV has been used for the determination of organophosphorus insecticides, as was demonstrated by Supin and Budnikov (1973) for the determination of phthalophos and benzophosphate in apples. A 200 g amount of apple was extracted with acetone, the extract evaporated and the final residue treated with alkali to produce dimethyl and diethyl thiophosphate, respectively. These compounds could be determined by CSV down to a detection limit of 0.2 $\mu g\ kg^{-1}$.

Davidek and Seifert (1971) developed an enzymic polarographic method for the determination of the organophosphorus pesticide intration. The method is based on the inhibition of antichlolinesterase activity by the organophosphorus

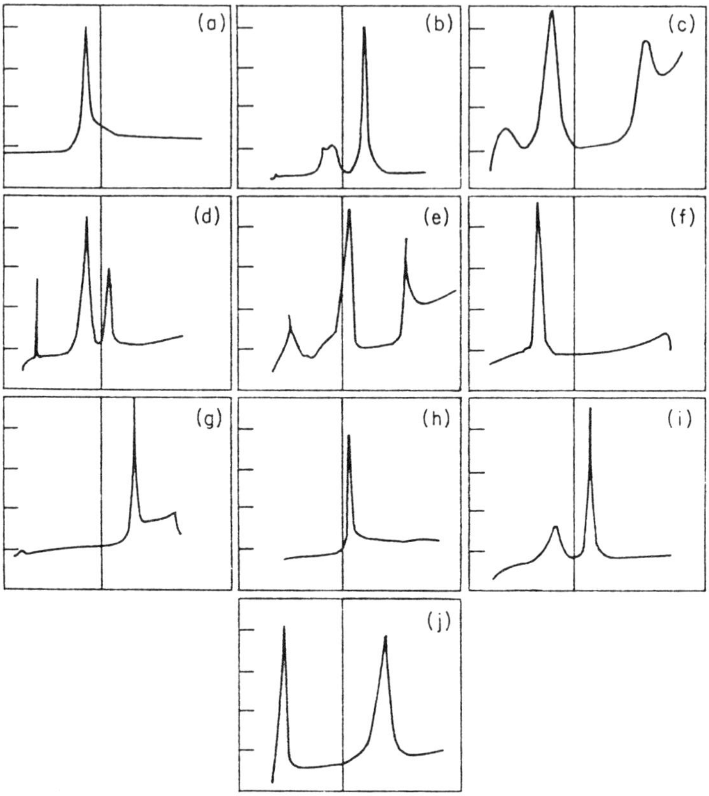

Figure 37. Linear-sweep voltammograms of several important organophosphorus pesticides (Nangniot, 1964) (current ranges given in parentheses). (a) 3.30×10^{-4} M chlorthion in 0.2 M KCl (0.5 μA); (b) 2.30×10^{-4} M prothoate in 0.2 M NaOH (0.2 μA); (c) 1.29×10^{-4} M methyl azinphos in 0.2 M NaOH (0.02 μA): (d) 2.82×10^{-4} M demeton-*O*-sulphoxide in 0.2 M NaOH (0.03 μA): (e) 2.21×10^{-4} M malathion in 0.2 M KCl (0.02 μA): (f) 1.22×10^{-4} M menazon in 0.2 M NaOH (0.08 μA): (g) 1.70×10^{-4} M phenkapton in 0.2 M KCl (0.03 μA): (h) 2.16×10^{-4} M parathion in 0.2 M NaOH (0.5 μA): (i) 1.74×10^{-4} M dimethoate in 0.2 M NaOH (0.2 μA), (j) 1.93×10^{-4} M demeton in 0.2 M NaOH (0.02 μA). Reproduced by permission of Elsevier Science Publishers from *Anal. Chim. Acta*, **31** (1964) 166

moiety. Unreacted enzyme is then incubated with β-naphthyl acetate and the β-naphthol liberated is measured by polarography, following nitrosation. The method was applied to the determination of intration in lettuce, cabbage, cherries and tomatoes. Naturally occurring enzymes, which would be capable of hydrolysing β-naphthyl acetate, were removed by precipitation with ethanol and subsequent centrifugation. No interference was observed in the presence of carotenes, xanthophyll, chlorophyll or anthocyanidines.

Nitro-containing organophosphorus compounds such as ethyl parathion (**XLIX**) have been determined in vegetables with electrochemical detection after

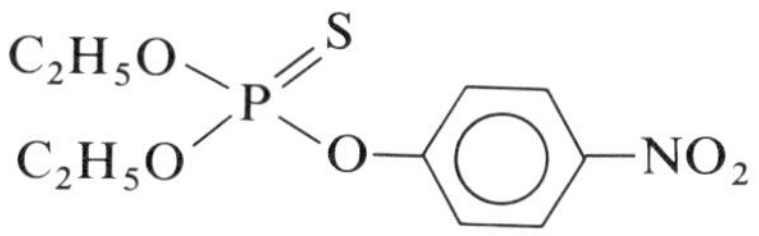

(XLIX)

liquid chromatography (Koen and Huber, 1970; Clark *et al.* (1985). The method of Clark *et al.* (1985) involved reversed-phase HPLC with series UV and electrochemical detection at – 0.85 V. Sample preparation techniques were developed which avoided the usual preliminary column fractionations and which allowed ethyl parathion (**XLIX**) and methyl parathion to be recovered with an average of 95 % at concentrations less than 50 ng g^{-1} for plant material. Relative standard deviations of 5% were obtained using five different plant samples. Concentrations less than 10 ng ml^{-1} were readily measured in water samples using a column concentration procedure. The selectivity of electrochemical detection made it unnecessary to resolve the electroactive plant components chromatographically from the pesticides and allowed rapid determination. Series detection proved useful in distinguishing various pesticides and in comparing the characteristics of the two detectors. The pesticides alachlor, atrazine, carbaryl, carbofuran, chloropyrifos, diazinon dyfonate and lannate did not give electrochemical detection under the conditions of the experiment.

P. Hernandez *et al.* (1990) recently published an AdSV method for the determination of azinphos-methyl (Guthion). The LOD was 0.2 ng ml^{-1} after 300 s accumulation at an HMDE and the method was applied to spiked river water.

7

Voltammetric Analysis of Organometallics

7.1 ORGANOMERCURY SPECIES

Organomercurials have been used as fungicides, pesticides and medicinals for most of this century. By the 1950s, however, the toxicological effects of mercury were becoming apparent, e.g. wildlife poisoning caused by eating mercury-treated seeds and methylmercury contamination of fish and shell fish in Minamata Bay, Japan, resulting in severe poisoning of the local population. The latter case involved inorganic mercury being converted by aquatic microorganisms into methylmercury, which causes irreversible damage to the central nervous system.

Brezina and Zuman (1958) reviewed the polarographic behaviour of the now rarely used organomercury pharmaceuticals. Phenylmercury derivatives such as ethylmercurithiosalicylate and *p*-chloromercuribenzoate are reduced in two waves. The first wave, corresponding to the formation of the organomercury radical, shows some complications from adsorption which can be overcome by the addition of a surface-active agent such as gelatin or Triton X-100. The best defined waves for this class of compound are found obtained with Britton–Robinson buffer (pH 9.0).

Hopes (1966) described a method for the polarographic determination of phenylmercury fungicides. The sample was digested with hydrochloric acid and the organomercurial extracted with chloroform. The chloroform was evaporated and the residue dissolved in an aqueous pH 9 buffer prior to polarographic analysis.

Fleet and Jee (1970) studied the free-radical reduction mechanism of organomercury compounds and summarised it as follows:

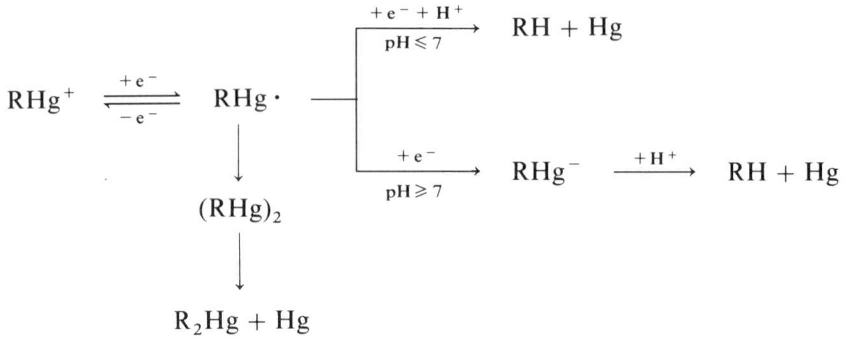

Fleet and Jee (1969) determined Hg(II)–alkene addition compounds by a.c. polarography in the solvent methanol–water (9 + 1) with 0.1 M NaOH or $NaNO_3$ as the supporting electrolyte. The method was proposed for the determination of a wide range of alkenes and could be used for the analysis of mixtures in view of the variation in reaction rates. The sample of alkene (0.1–2 meq.) dissolved in methanol was placed in a 10 ml volumetric flask. An excess of methanolic mercury(II) acetate solution was added so that the final concentration would be 2×10^{-4}–3×10^{-4} M. After a suitable time had elapsed for the formation of the organomercury compound, 1 ml of 1 M NaOH was added as supporting electrolyte and the solution made up to 10 ml with methanol. This solution was deaerated and subjected to a.c. polarography. The alkenes, allylacetone, vinyl acetate, *N*-vinylcarbazole, *n*-butyl vinyl ether, isobutyl vinyl ether and 2-chloroethyl vinyl ether react to form derivatives in 2 min at room temperature and give a.c. peak potentials in the range − 0.41 to − 0.55 V (vs SCE) and peak half-widths in the range 80–129 mV.

Heaton and Laitinen (1974) showed a linear dependence of DPP current on concentration for the reduction of the environmentally significant entity CH_3Hg^+ in the range 10^{-4}–10^{-7} M. MacCrehan *et al.* (1977) combined HPLC with reductive ED to separate the organomercurials CH_3Hg^+, $C_2H_5Hg^+$, and $PhHg^+$ as their zero-charged 2-mercaptoethanol complexes on a Spherosorb ODS 5 μm column, 250 × 4.6 mm i.d., with a mobile phase of methanol–water (40 + 60) containing 0.06 M NH_4OAc (pH 5.5), and 5×10^{-4} M 2-mercaptoethanol, flow rate 1.0 ml min^{-1}, and a detector potential set at − 0.90 V (Fig. 38). Linear calibration graphs can be obtained over a wide concentration range and the detection limits are in the low nanogram range.

Such as HPLC–ED technique is of significant value in speciation studies of organometallics. Not only is the retention time characteristic of a particular organometallic species, as shown in Fig. 39 (MacCrehan *et al.* 1977), but also different organometallics reduce at different potentials (Table 29), thus enabling detector potential control to be used for characterisation purposes. It should be noted that the background currents increase significantly beyond − 1.0 V (vs Ag/AgCl) (Fig. 40).

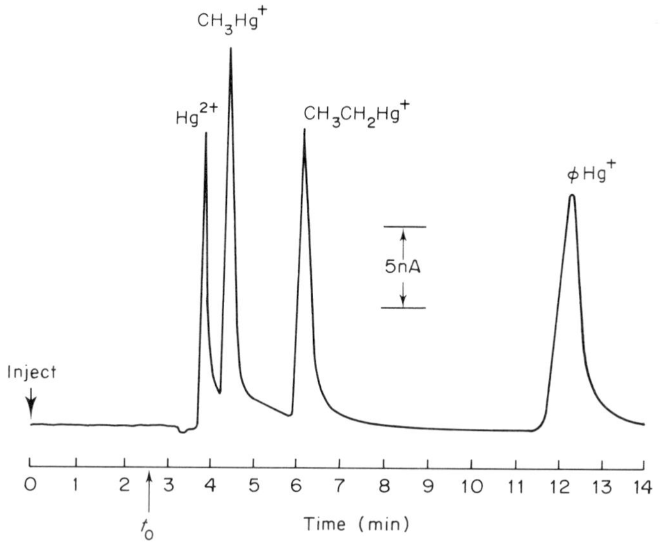

Figure 38. Separation of some organomercurials as their 2-mercaptoethanol complexes. Conditions: column, Spherosorb ODS, 5 μm (250 × 4.6 mm i.d.), mobile phase, methanol–water (40 + 60) containing 0.06 mol l^{-1} NH_4OAc (pH 5.5) and 5×10^{-4} mol l^{-1} 2-mercaptoethanol; flow rate, 1.0 ml min^{-1}; detector potential, − 0.90 V; current sensitivity, 50 nA V^{-1}, background offset, 12.0 nA; sample amounts, 130 ng Hg^{2+}, 140 ng CH_3Hg^+, 300 ng $CH_3CH_2Hg^+$, 370 ng $PhHg^+$. Reprinted from MacCrehan *et al.*, *Analytical Letters*, **10** (1977) 1175 by courtesy of Marcel Dekker, Inc.

7.2 ORGANOTIN SPECIES

Organotin compounds have found extensive application as wood preservatives, stabilisers for PVC plastics, fungicides and antifouling paint additives. Alkyltin compounds are more toxic than the aryl compounds and the trisubstituted derivatives more toxic than the di- or tetrasubstituted derivatives. Environmentally, organotin compounds have an advantage over alternative organometallics in that the inorganic tin formed on degradation is essentially non-toxic.

The pesticide fentin, which contains the triphenyltin species Ph_3Sn^+, will reduce in a reversible $1e^-$ process to give the radical $Ph_3Sn\cdot$, which can adsorb on the mercury surface. Further reduction and protonation to give Ph_3SnH or dimerisation to $Ph_3SnSnPh_3$ can then take place. Booth and Fleet (1970) recommended that the first adsorption peak, observed on the application of DPP and corresponding to the formation of the organotin free radical, be used for trace analytical purposes. In addition, they showed that fentin residues can be determined in the concentration range 10^{-7}–10^{-8} M by ASV using the adsorption of the species $Ph_3Sn\cdot$ as the accumulation step. Although there was a slight loss of radical during this step due to dimerisation or deactivation, the method was found to be precise. It was found necessary to control the length of

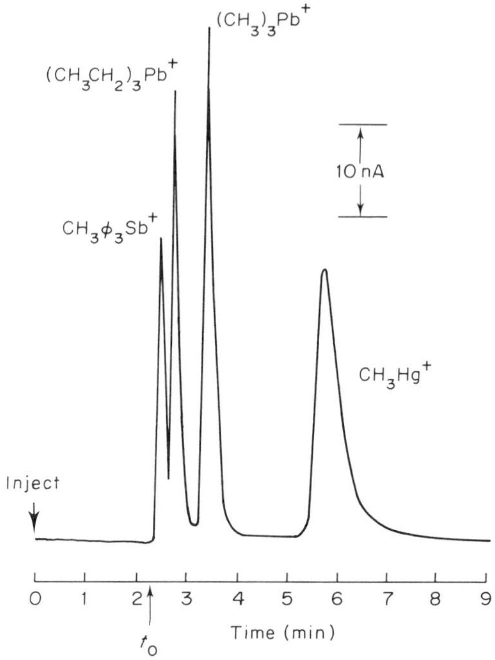

Figure 39. Separation of different organometals. Conditions: column, Lichrosorb NH_2, 10 μm (250 × 4.6 mm i.d.) mobile phase, methanol–water (40 + 60) containing 0.06 mol l^{-1} NH_4OAc (pH 5.5); flow rate, 1.0 ml min^{-1}; detector potential, − 1.00 V; current sensitivity, 100 nA V^{-1}; background offset, 6.0 nA; sample amounts, 180 ng $CH_3Ph_3Sb^+$, 430 ng $(CH_3CH_2)_3Pb^+$, 170 ng $(CH_3)_3Pb^+$, 400 ng CH_3Hg^+. Reprinted from MacCrehan *et al.*, *Analytical Letters*, **10** (1977) 1175 by courtesy of Marcel Dekker, Inc.

Table 29. Peak potentials (E_p) for the first reduction wave of selected organometals. Reprinted from *Analytical Letters*, **10** (1977) 1175 by courtesy of Marcel Dekker, Inc.

Compound	E_p (V)	Compound	E_p (V)
CH_3Hg^+	−0.35	$(CH_3)_3Sb^{2+}$	−0.94
$CH_3CH_2Hg^+$	−0.33	Ph_3Sb^{2+}	−0.77
$PhHg^+$	−0.65	Ph_3Sb	−0.49
		$CH_3Ph_3Sb^+$	−0.80
CH_3Sn^{3+}	−1.07	Ph_4Sb^+	−0.89
$(CH_3)_2Sn^{2+}$	−0.87	$(CH_3)_3Pb^+$	−0.57
$(CH_3)_3Sn^+$	−1.14	$(CH_3CH_2)_3Pb^+$	−0.49
Ph_2Tl^+	−0.44	Ph_3Pb^+	−0.50

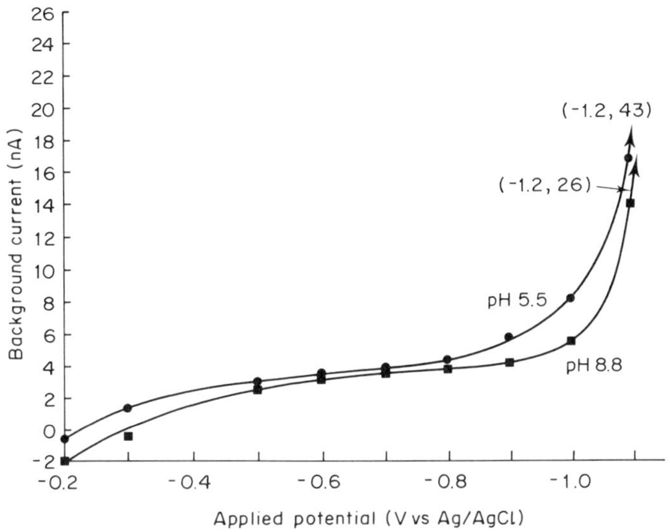

Figure 40. Background residual current as a function of applied potential in methanol–water (60 + 40) containing 0.04 mol l^{-1} NH_4OAc. Reprinted from MacCrehan *et al.*, *Analytical Letters*, **10** (1977) 1175 by courtesy of Marcel Dekker, Inc.

this accumulation step so that less than a monolayer of the free radical was deposited on the electrode surface. When more than a monolayer was deposited, two distinct stripping peaks were obtained, corresponding to the removal of the bulk and surface monolayer. Using this procedure, levels of fentin in potatoes were determined following extraction of the macerated sample with acetonitrile and direct measurement of the aqueous organic extract (Table 30). The resulting method had an approximately 100 times lower LOD than conventional spectrophotometric procedures.

Polarographic determination of dialkyltin compounds can be carried out over the concenration range 10^{-4}–10^{-9} M by DPP using acetate buffer (pH 7) containing 80% (v/v) ethanol and utilising the first reduction peak due to $Bu_2Sn^{2+} \underset{-e^-}{\overset{+e^-}{\rightleftharpoons}} Bu_2Sn^{+}\cdot$. Using the same supporting electrolyte, monoorganotin compounds could be determined down to 5×10^{-7} M. Some resolution of mixtures of organotin species could be achieved polarographically as shown in Fig. 41 (Fouzder and Fleet, 1979). Further improvement in selectivity was possible by recourse to HPLC–ED with a glassy carbon wall jet cell, e.g. separation of dibutyltin compounds from their mixture with tributyltin oxide. Walton *et al.* (1991) recently used HPLC with laser-excited atomic fluorescence in a flame as a detector to resolve satisfactorily mixtures of tributyl-, trimethyl- and triethyltin for environmental applications and Ebdon (1991) recommended the use of HPLC–ICP–MS for the determination of inorganic and organic tin species in environmentally relevant samples.

Table 30. Samples of potatoes analysed for fentin (Booth and Fleet, 1970)

	Skin		$\frac{1}{4}$-in layer		Bulk		
Sample	$\mu g\ ml^{-1}$	Potato (%)	$\mu g\ ml^{-1}$	Potato (%)	$\mu g\ ml^{-1}$	Potato (%)	Total ($\mu g\ ml^{-1}$
1	0.03	8	0.01	70	0.001	22	0.008
2	0.05	12	0.001	47	0.001	41	0.007
3	ND[a]	8	0.01	60	ND	32	0.005
4	ND	8	0.01	51	ND	41	0.005
5	0.001	8	0.002	64	ND	28	0.001

[a]ND, none detected.

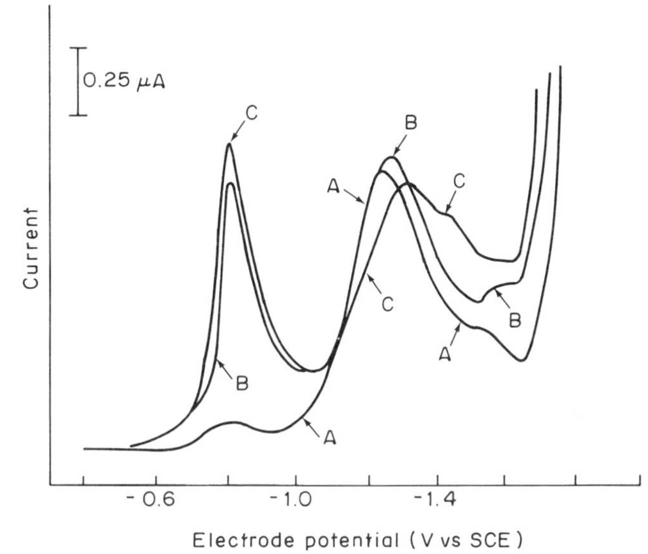

Figure 41. Differential-pulse polarogram of a 2.23×10^{-5} mol l^{-1} phenyltin trichloride solution in acetate buffer (pH 7.0) containing 80% (v/v) ethanol. (A) Pure phenyltin trichloride solution; (B) in the presence of diphenyltin dichloride (2.91×10^{-5} M); (C) in the presence of di- and triphenyltin compounds (2.91×10^{-5} and 1.43×10^{-5} M respectively. Reproduced by permission of Academic Press from Smyth, *Polarography of Molecules of Biological Significance*, 1979

7.3 ORGANOLEAD SPECIES

Organic derivatives of lead have widespread industrial applications, primarily as antiknock additives in petrol and also as polymerisation catalysts, antifouling agents and alkylating agents for the production of organomercurial fungicides. Tetralkyllead species are toxic owing to the formation of the trialkyllead cation, which inhibits oxidative phoshorylation, while the toxicity of dialkyllead compounds is supposedly due to their ability to complex enzyme sulphydryl groups. Speciation/toxicological studies have assumed considerable importance owing to the considerable concern over environmental lead pollution.

In acetate buffer (pH 7) containing 50% ethanol, the species Ph_3Pb^+ showed two well defined DPP peaks which were diffusion controlled and linearly dependent on concentration over a wide range (e.g. 2×10^{-4}–5×10^{-7} M for the first peak corresponding to $Ph_3Pb^+ \xrightarrow{+e^-} Ph_3Pb\cdot$). A similar supporting electrolyte could be used for the determination of dialkyllead compounds (Fouzder and Fleet, 1979).

An ASV method has been recommended for the determination of organolead in petroleum (Roschig and Matschiner, 1967). A solution of bromine in chloroform was used to decompose R_4Pb followed by extraction of Pb(II) into 0.1 M HNO_3. In this method, accumulation was carried out at − 0.80 V (vs SCE) for 1.5 min and Pb(II) determined at − 0.39 V (vs SCE). An LOD of 8 parts of Pb per 10^{13} with a relative standard deviation of 7% was claimed.

MacCrehan *et al.* (1977) used reversed-phase HPLC–reductive ED to detect 0.1 ng (corresponding to 2×10^{-8} M or 5 ppb) of the species $(CH_3)_3Pb^+$ with linear calibration in the concentration range 10^{-3}–10^{-7} M. They stated that some form of sample concentration would be necessary for the determination of $(CH_3)_3Pb^+$ in 'real' water samples.

7.4 ORGANOARSENIC SPECIES

The organic compounds of arsenic have found a wide range of applications as additives to animal feeding stuffs, treatment of African trypanosomiases, pesticides, defoliants and chemical warfare agents. Many of the papers published on the determination of organoarsenic compounds deal with elemental analysis. This can have the disadvantage of loss of volatile arsenic during the decompositions procedure and incomplete decomposition of the sample. In addition, elemental analysis does not differentiate between the different oxidation states, which vary considerably in toxicity.

The most widespread technique for the determination of arsanilic acid (4-aminophenylarsonic acid, **L**) the most commonly used arsonic acid in animal feeding stuffs, is diazotisation of the amino group followed by coupling with an agent such as *N*-naphthylethylenediamine to form a highly coloured dyestuff which is measured spectrophotometrically (Malaiyandi *et al.* 1969). This method has a precision of ca 2%, is easily automated and can be used to determine selectively aminophenylarsonic acids in the presence of other arsonic acids. It again has the disadvantages that it does not differentiate between oxidation states and is not a direct measure of the toxicity of a sample since the therapeutically active arsonic group is being determined indirectly with any other amino compounds present interfering. Raney nickel can be used to break the C—As bond by reduction, with the organic residue being determined by GC (George and Morrison, 1971). This method has the great advantage of selective analysis of different substituted phenylarsonic acids, but again does not differentiate between the oxidation states.

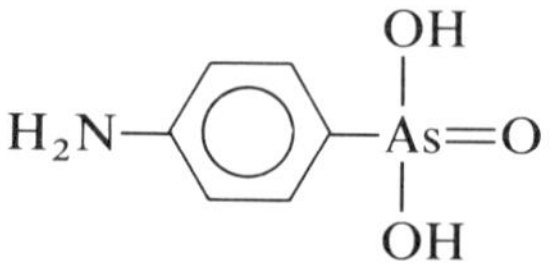

(**L**)

The derivatives of phenylarsonic acid give rise to a single, well defined diffusion-controlled wave in 0.1 M HCl (Watson and Svehla, 1975a). The $E_{1/2}$

values are too close to resolve mixtures, except when the substituent itself is electroactive, e.g. containing an NO_2 group. The lower oxidation state, phenyl arsenoxide (**LI**), can be differentiated from **L** in 0.1 M HCl (Watson and Svehla, 1975b).

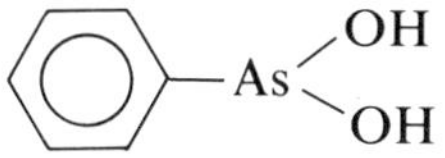

(**LI**)

During formation of the carbon—arsenic bond in the manufacture of an arsonic acid from inorganic arsenic, a certain percentage of di- and trisubstitution occurs and so the diphenylarsinic acid (**LII**) and triphenylarsine oxide (**LIII**) are often found as impurities in the phenyl arsonic acid. Quantitative detection of these in the arsonic acid is often difficult. Again, polarography offers a possibility. The phenylarsonic acids (Watson and Svehla, 1975a) are not electroactive above pH 3 whereas the diphenylarsinic acids (Watson, and Svehla, 1974) remain electroactive until their pK_1 value about pH 6 is reached, above which triphenylarsine oxide (Watson and Svehla, 1975c) remains active in the remainder of the pH range. Table 31 gives an example of analyses by Maruyama and Furuya (1957) of bis(aminophenyl)arsinic acid in 4-aminophenylarsonic acid. Their error of 4–5% could be improved by the use of the sampled d.c. technique.

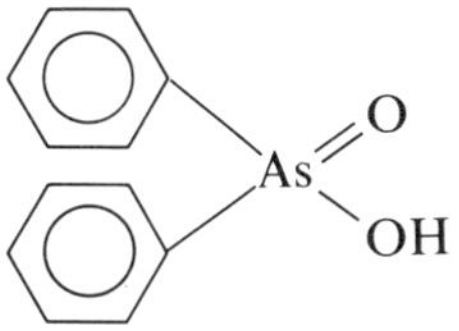

(**LII**)

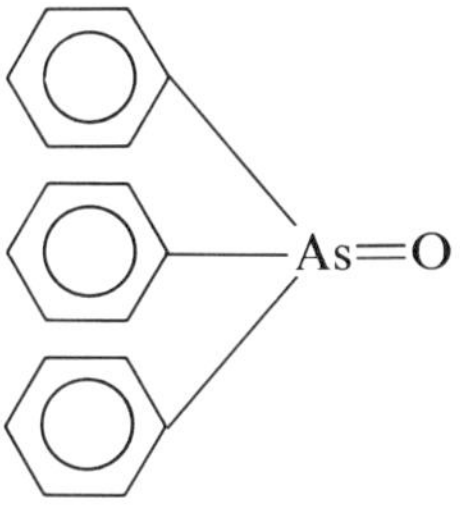

(**LIII**)

Table 31. Determination of bis(aminophenyl)arsinic acid in 4-arsanilic acid (4-aminophenylarsonic acid)

Sample No.	Calculated value (%) 4-Arsanilic acid	Calculated value (%) Bis(aminophenyl)-arsinic acid	Observed value for bis(aminophenyl)-arsinic acid (%)
1	85.0	15.0	14.0
2	70.0	30.0	30.1
3	60.0	40.0	41.7
4	40.0	60.0	56.0
5	20.0	80.0	76.0

Table 32. Determination of selected organoarsenic compounds in the presence of each other by DPV. Modulation amplitude, 50 mV; scan rate, 20 mV s^{-1}; pre-electrolysis time, 0 min

Solution	Supporting electrolyte	Initial potential (V)	Interferent added	Recovery (%)		
				$E_p - 0.4$ V	$E_p - 0.8$ V	
10^{-5} M phenylarsine oxide	pH 4	−0.2	None	100	100	
			10^{-5} M As(III)	100	100	
			10^{-5} M As(V)	100	100	
			10^{-5} M triphenylarsine	100	100	
			10^{-5} M triphenylarsine oxide	100	100	
			10^{-5} M arsenazo	100	Very high	
				$E_p - 0.17$ V	$E_p - 0.28$ V	$E_p - 0.60$ V
10^{-5} M arsenazo	pH 3	+0.2	10^{-5} M triphenylarsine	100	100	100
			10^{-5} M triphenylarsine oxide	100	214	Very high
			10^{-5} M As(III)	100	90	100
			10^{-5} M As(V)	100	100	100
				E_p 0.9 V	$E_p - 1.13$ V	
10^{-5} M triphenylarsine oxide	pH 4	−0.2	10^{-5} M arsenazo	91	100	
		−0.4	10^{-5} M arsenazo	97	100	
		−0.2	10^{-5} M As(V)	100	100	
		−0.2	10^{-5} M As(III)	100	Very high	
		−0.2	10^{-5} M triphenylarsine	100	100	
		−0.2	10^{-5} M phenylarsine oxide	Very high	Very high	

Jan and Smyth (1984b) investigated the differential-pulse voltammetric behaviour of selected inorganic and organoarsenic compounds at the HMDE, the main aim being to determine a particular arsenic compound whether it be in the + 3 or + 5 oxidation state, inorganic or organic, and in the presence of other forms using voltammetric techniques. The results are summarised in Table 32. In addition, it was found difficult to determine As(III) by this technique when interfering organoarsenic molecules were present in the mixture.

These purely electroanalytical techniques are therefore of limited selectivity in speciation studies of arsenic and it is necessary to link, say, ED with HPLC as MacCrehan *et al.* (1977) did for mercury (Fig. 38) and other organometallic species (Fig. 39, Table 29). Such ED subsequent to HPLC should provide complementary information to ICP–MS detection following HPLC in speciation studies of arsenic.

8

Voltammetric Analysis of Selected Inorganic Molecules

Rocklin and Johnson (1983) determined CN^- and other anions by ion chromatography (IC) and electrochemical detection with a silver working electrode. This ED was used in conjunction with a conductivity detector to determine ions detectable by either method. The detector was placed between the separator and suppressor columns of the ion chromatograph, since the suppressors's purpose is to lower the conducitivity by decreasing the concentration of the supporting electrolyte. The separation of a ten anion standard is shown in Fig. 42. HS^-, CN^-, Br^- and $SO_3{}^{2-}$ were detected by the silver electrode whereas $NO_2{}^-$, $NO_3{}^-$, $HPO_4{}^{2-}$ and $SO_4{}^{2-}$ produced no response. Owing to the low dissociation of H_2S and HCN following protonation by the suppressor column, they were not detected by the conductivity detector. The cyanide in inorganic cyanides can be present as both complexed and free cyanide. Table 33 lists the percentage of total cyanide detected by IC–ED for selected metal cyanides. This suggests that the complex cyanides can be grouped into three categories, depending on the cumulative formation constant and stability of the complex. Category 1 includes the weakly complexed and labile cyanides $Cd(CN)_4{}^{2-}$ ($\log \beta_4 = 18.78$) and $Zn(CN)_4{}^{2-}$ ($\log \beta_4 = 16.7$). These complexes dissociate completely under the chromatographic conditions used, the cyanide being indistinguishable from free cyanide. Category 2 includes the moderately strong cyanide complexes $Ni(CN)_4{}^{2-}$ ($\log \beta_4 = 31.3$) and $Cu(CN)_4{}^{3-}$ ($\log \beta_4 = 30.3$). Although the complexes are labile, they are retained on the column and slowly dissociate during the chromatography, as evidenced by tailing in Fig. 43 for $Cu(CN)_3{}^{2-}$. Category 3 includes those cyanides which are inert and therefore totally undissociated such as $Au(CN)_2{}^-$ ($\log \beta_2 = 38.3$), $Fe(CN)_6{}^{3-}$ ($\log \beta_6 = 42$) and $Co(CN)_6{}^{3-}$ ($\log \beta_6 = 64$). No free cyanide was detected for these complexes. It is therefore possible to analyse samples for free cyanide (or weakly complexed cyanide) and strongly complexed cyanide by direct injection. The determination of total cyanide requires distillation of the sample with caustic trapping.

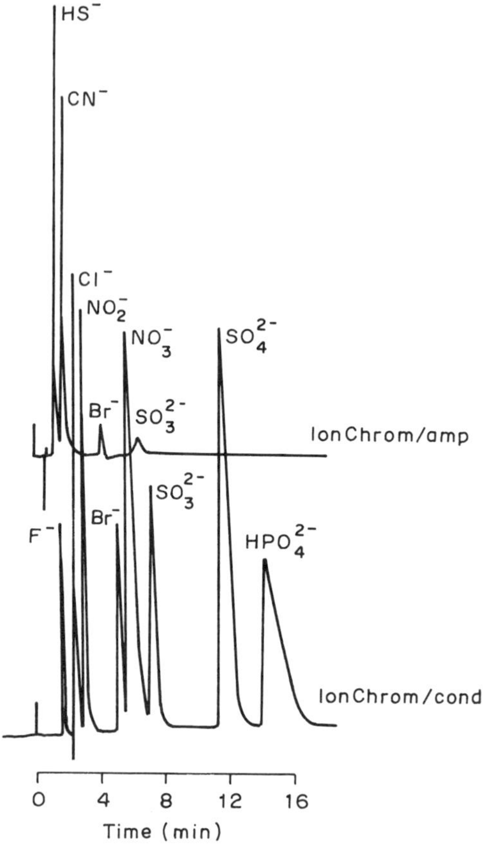

Figure 42. Simultaneous analysis by using electrochemical and conductivity detection. Concentrations are 300 ppb sulphide, 500 ppb cyanide, 1 ppm fluoride, 4 ppm chloride, 10 ppm nitrite, 10 ppm bromide, 25 ppm nitrate, 30 ppm sulphite, 25 ppm sulphate and 50 ppm phosphate. Reprinted with permission from *Analytical Chemistry*, **55**, 4. Copyright (1983) American Chemical Society

Table 33. Percentage of total cyanide in metal complexes determined as 'free' cyanide. Reprinted with permission from Takano *et al.*, *Analytical Chemistry*, **56**, 1594. Copyright (1983) American Chemical Society

Metal complex	Log β_f	%
$Cd(CN)_4^{2-}$	18.8	102
$Zn(CN)_4^{2-}$	16.7	102
$Ni(CN)_4^{2-}$	31.3	81
$Cu(CN)_4^{3-}$	30.3	52
$Cu(CN)_3^{2-}$	28.6	42
$Cu(CN)_2^{-}$	24.0	38
$Au(CN)_2^{-}$	38.3	0
$Fe(CN)_6^{3-}$	42	0
$Co(CN)_6^{3-}$	64	0

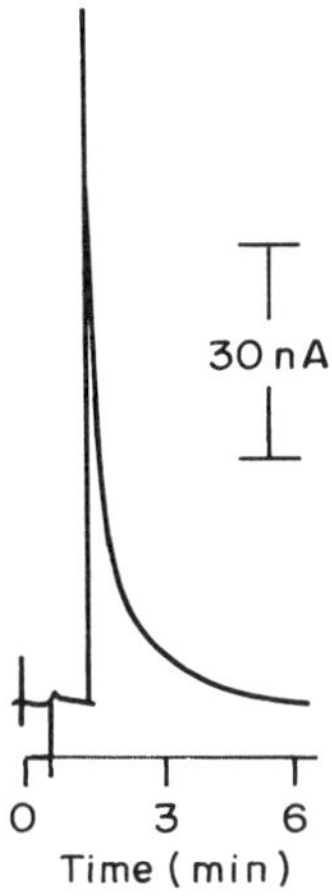

Figure 43. Chromatogram of 1.0×10^{-5} M $Cu(CN)_3^{2-}$. Reprinted with permission from Allison and Sharp *Analytical Chemistry*, **55**, 4. Copyright (1983) American Chemical Society

The presence of chloramines such as NH_2Cl, $NHCl_2$ and NCl_3, formed during the chlorination of water, may give rise to taste and odour problems in drinking water and therefore they have to be monitored at the low $\mu g\ ml^{-1}$ level. Ge *et al.* (1990) used reversed-phase HPLC with ED at a glassy carbon electrode to monitor these chloramines. The electrochemical detector has a lower LOD than the UV detector for these species, particularly for dichloramine.

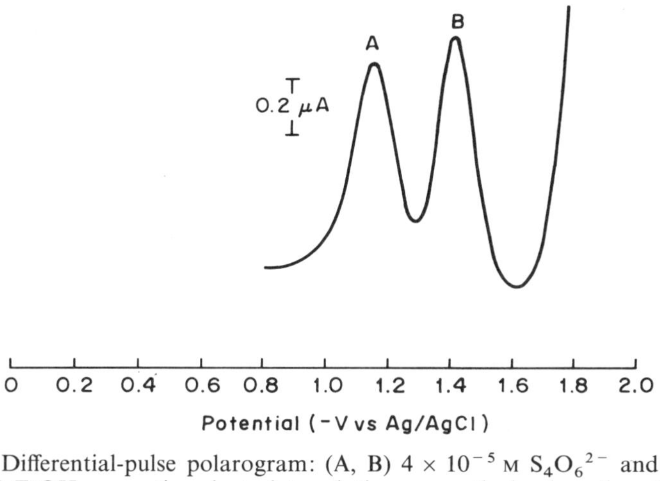

Figure 44. Differential-pulse polarogram: (A, B) 4×10^{-5} M $S_4O_6^{2-}$ and $S_3O_6^{2-}$ in CsCl–TMC–EtOH supporting electrolyte solution, respectively; drop time, 0.5 s; modulation amplitude, 50 mV; scan rate, 5 mV s^{-1}. Reprinted with permission from *Analytical Chemistry*, **56**, 1594. Copyright (1984) American Chemical Society

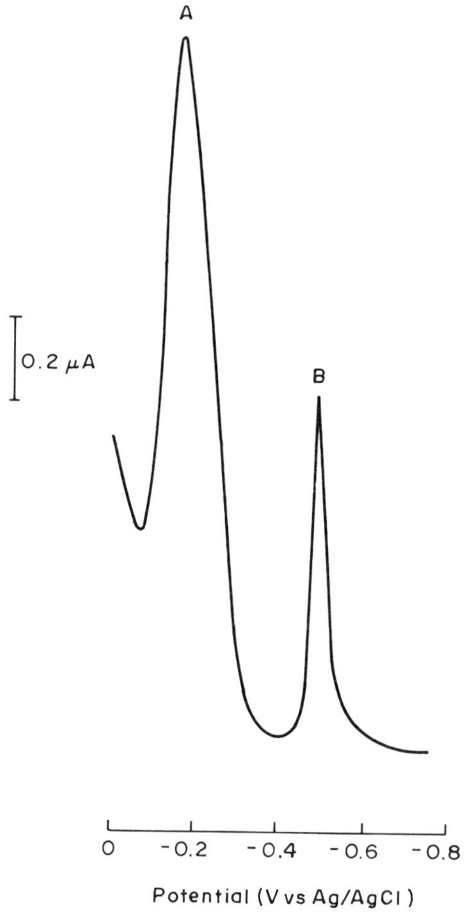

Figure 45. Differential-pulse polarogram of pentathionate with tetrathionate: supporting electrolyte, 1 M K_2HPO_4–Na_2HPO_4 (pH 6.5); drop time, 0.5 s; modulation amplitude, 50 mV; scan rate, 5 mV s^{-1}; mercury drop size, medium: (A) 5×10^{-5} M $S_4O_6^{2-}$, (B) 5×10^{-6} M $S_5O_6^{2-}$. Reprinted with permission from *Analytical Chemistry*, **56**, 1594. Copyright (1984) American Chemical Society

Anion-exchange HPLC of polythionates and thiosulphate followed by DPP has been shown to be a convenient method for the determination of these sulphur species in mixtures (Takano *et al.*, 1984); HPLC can initially separate the mixture into three groups, thiosulphate, tri- + tetra- + pentathionate and hexathionate. Tri- and tetrathionates are the major polythionate ions in soils, lake sediments, mining effluents and pyrite-leaching solutions and these ions can be resolved by DPP in the supporting electrolyte, (0.5 M CsCl–0.2 M tetramethy-

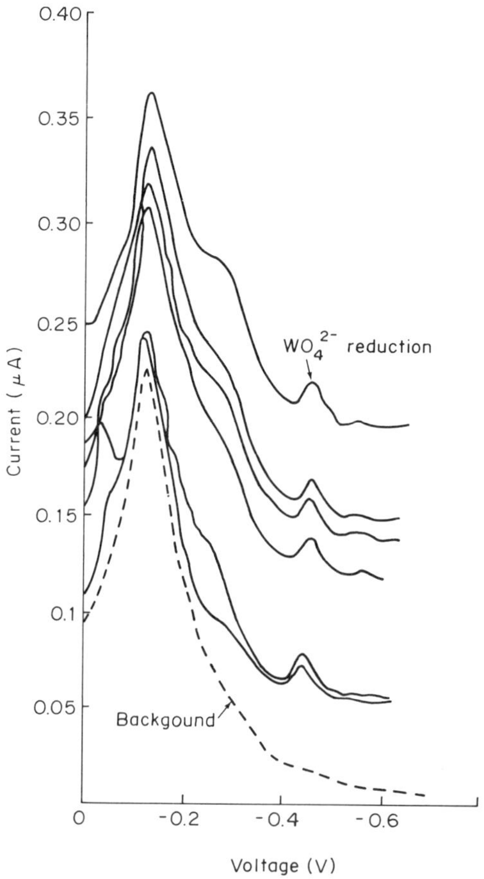

Figure 46. Repetitive scan at a concentration (4×10^{-8} M WO_4^{2-}) near the detection limit for WO_4^{2-} by a DPV method in 0.06 M phosphate buffer solution (the curves are off-set). Reprinted by permission of the Royal Society of Chemistry

lammonium chloride)–ethanol (1:1, v/v) (Fig. 44). If pentathionate is present in the sample at more than a tenth of the concentration of tetrathionate, the supporting electrolyte is changed to phosphate buffer (pH 6.5) for determination of pentathionate (Fig. 45). $S_2O_3^{2-}$ has also been measured in human urine and plasma by a similar technique (Kawanishi *et al.* 1984).

Jan and Smyth (1984a) studied the DPV behaviour and determination of the oxyanions CrO_4^{2-}, WO_4^{2-}, MoO_4^{2-} and VO_3^- at the HMDE. LODs of 2.7×10^{-6} M for CrO_4^{2-}, 7.7×10^{-9} M for WO_4^{2-} and 1.4×10^{-5} M for VO_3^- were found. Figure 46 shows repetitive scans of 4×10^{-8} M WO_4^{2-} in 0.06 M

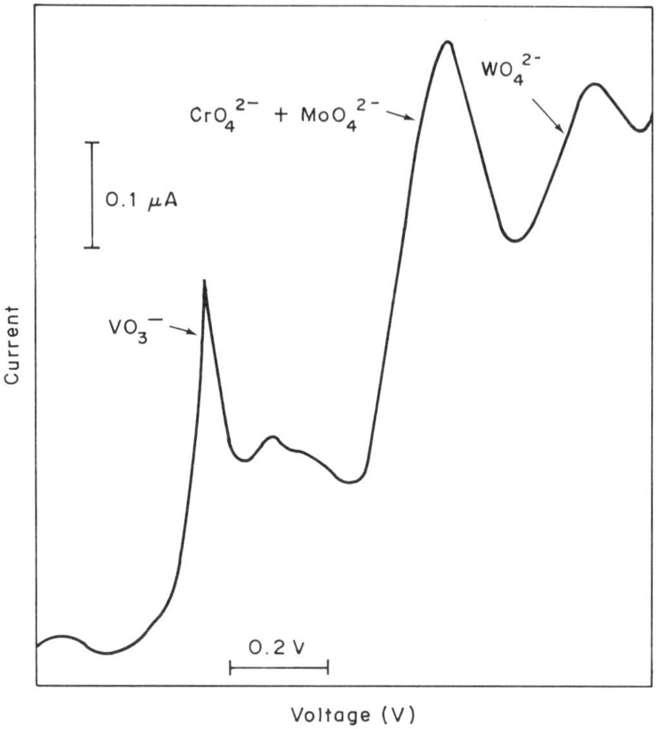

Figure 47. Resolution and identification of an equimolar (10^{-5} M) mixture of oxyanions by DPV in pH 5 phosphate buffer solution. Reprinted by permission of the Royal Society of Chemistry

phosphate buffer solution and Fig. 47 the voltammetric resolution of a mixture of these oxyanions. Molybdenum can also be determined by DPAdSV in phosphate buffer (pH 2) utilising the adorption of 12-molybdophosphoric acid at the HMDE (Fogg and Alonso, 1988). Using a 1 min accumulation time, calibration graphs were linear up to 7×10^{-7} M Mo and an LOD of $< 10^{-9}$ M can be expected. Figure 48 shows the DPAdSV for different Mo(VI) concentrations and Fig. 49 the DPAdSV for 5.6×10^{-9} M Mo(VI) in phosphate buffer for different accumulation times.

Lewis *et al.* (1983) used HPLC–ED for the determination of pertechnetate (TcO_4^-) ($^{99m}TcO_4^-$ + $^{99}TcO_4^-$) in ^{99}Mo–^{99m}Tc generator eluents in the concentration range 5×10^{-8}–5×10^{-6} M . No sample treatment was found necessary since other components were separated by the amino-bonded column. The SMDE was used together with rigorous deoxygenation procedures. Figure 50 shows a chromatogram of a typical generator sample and the LOD was found to be 4.2×10^{-8} M TcO_4^- with a relative standard deviation of case 2%.

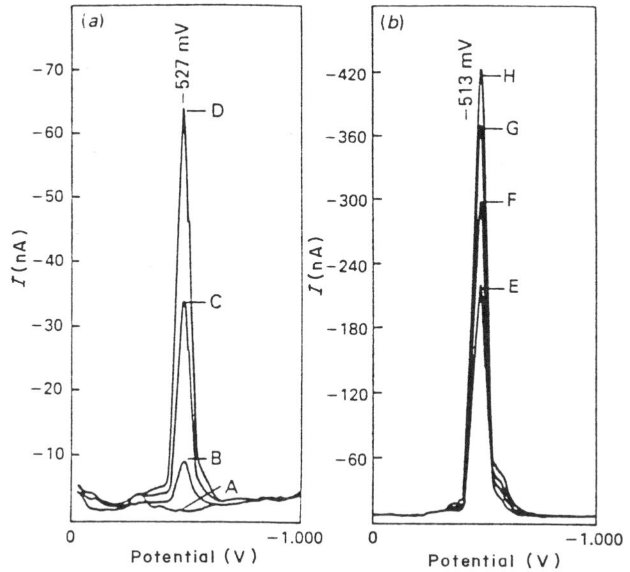

Figure 48. Differential-pulse adsorptive stripping voltammograms for different molybdenum(VI) concentrations. Accumulation potential, -100 mV. Accumulation time, 1 min. Molybdenum concentration: (A) 0; (B) 1.1×10^{-7}; (C) 2.2×10^{-7}; (D) 3.3×10^{-7}; (E) 4.0×10^{-7}; (F) 5.1×10^{-7}; (G) 6.2×10^{-7}; and (H) 7.4×10^{-7} M. Reprinted by permission of the Royal Society of Chemistry

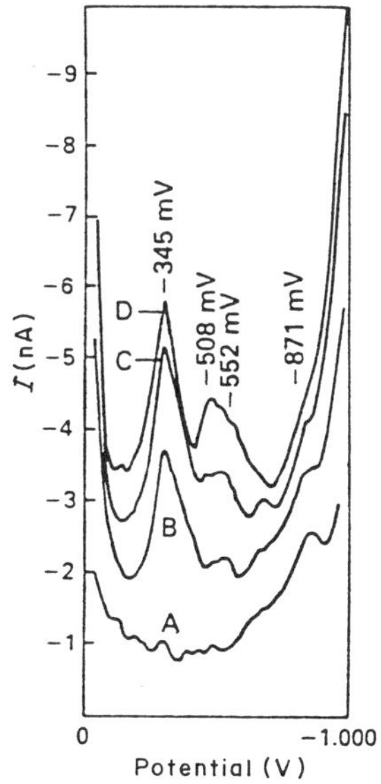

Figure 49. Differential-pulse adsorptive stripping voltammograms for molybdenum (5.6×10^{-9} M) in phosphate buffer (pH 2.0). Accumulation time: (A) 0; (B) 2; (C) 4; and (D) 6 min. Reprinted by permission of the Royal Society of Chemistry

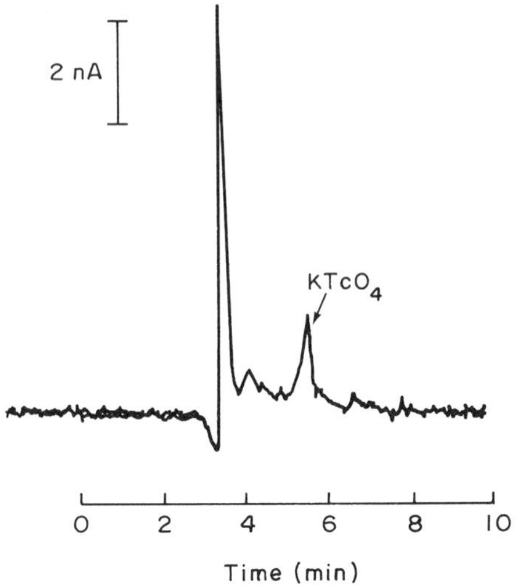

Figure 50. Chromatogram of a $KTcO_4$ sample from a ^{99}Mo–^{99m}Tc generator: mobile phase, 0.1 M phosphate buffer (pH 5.3); SMDE; sampled d.c. mode; $E_{applied} = -1.1$ V; 1 drop s^{-1}, flow rate, 1.0 ml min^{-1}. Reprinted with permission from *Analytical Chemistry*, **55**, 708. Copyright (1983) American Chemical Society

9

References

Afghan, B. K., Kulkarni, A. V., and Ryan, J. F. (1975). *Anal. Chem.*, **47**, 488.
Allison, L. A., and Shoup, R. E. (1983). *Anal. Chem.*, **55**, 8.
Alonso, R. M., Gallo, B., and Fogg, A. G. (1987). *Analyst*, **112**, 1611.
Altinoz, A., Temizer, A., and Beksac, S. (1990). *Analyst*, **115**, 873.
Amin, M., and Reusch, J. (1987). *Analyst*, **112**, 989.
Anderson, J. L., and Chesney, D. J. (1980). *Anal. Chem.*, **52**, 2156.
Anderson, J. L., Whiten, K. K., Brewster, J. D., Ou, T. Y., and Nonidez, W. K. (1985). *Anal. Chem.*, **57**, 1366.
Andrews, R. W., and King, R. M. (1990). *Anal. Chem.*, **62**, 2130.
Appelqvist, T., Marko-Vargo, G., Gorton, L., Torstensson, A., and Johansson G., (1985). *Anal. Chim. Acta*, **169**, 237.
Arcos, J., Vire, J.-C., El Jammal, A., and Patriarche, G. J., (1990). *Talanta*, **37**, 661.
Armentrout, D. N., and Cutie, S. S. (1980). *J. Chromatogr. Sci.*, **18**, 370.
Armentrout, D. N., McLean, J. D., and Long, N. W. (1979). *Anal. Chem.*, **51**, 1039.
Azewaka, T., Sano, A., Aoi, K., Sei, H., and Morita, Y. (1990). *J. Chromatogr.*, **530**, 47.
Baggesgaard-Rasmussen, H., Hahn, C., and Ilver, K. (1945). *Dan. Tidskr. Farm.* **19**, 41.
Becerro Dominguez, F., Gonzalez Diego, F., and Hernandez Mendez, J. (1990). *Talanta*, **37**, 655.
Beckett, A. H., Essien, E. E., and Smyth, W. F. (1974). *J. Pharm. Pharmacol.*, **26**, 399.
Beckett, A. H., Rahan, N. N., and Smyth, W. F. (1977). *Anal. Chim. Acta*, **92**, 353.
Belal, F., and Anderson, J. L. (1985). *Analyst*, **110**, 1493.
Berry, D. J. (1971). *Clin. Chim. Acta*, **32**, 235.
Bieder, A., and Brunel, P. (1971). *Annl. Pharm. Fr.*, **29**, 461.
Billedeau, S. M., Holder, C. L., and Getek, T. A., (1990). *J. Chromatogr.*, **534**, 151.
Birch, B. J., and Hart, J. P. (1979). In W. F. Smyth (ed.), *Polarography of Molecules of Biological Significance*, Academic Press, London, 209.
Bond, A. M. (1980). *Modern Polarographic Methods in Analytical Chemistry*, Marcel Dekker, New York.
Booth, M. D., and Fleet, B. (1970). *Anal. Chem.*, **42**, 825.
Bories, F. G., Peleran, J. C., Wal, J. M., and Corpet, D. E. (1983). *Drug Metabo. Dispos.*, **11**, 249.
Brand, M. J. D., and Fleet, B. (1968), *Analyst*, **93**, 498.
Brand, M. J. D., and Fleet, B. (1970), *Analyst*, **95**, 1023.
Bratin, K., Kissinger, P. T., Briner, R. C., and Bruntlett, C. S. (1981). *Anal. Chim. Acta*, **130**, 295.
Breyer, P. L., and Gilbert, B. P. (1987). *Anal. Chim. Acta*, **201**, 33.
Brezina, M., and Zuman, P. (1958), *Polarography in Medicine, Biochemistry and Pharmacy*, Interscience, New York, p. 527.

Bronstad, J. O., and Friestad, H. O., (1976). *Analyst*, **101**, 820.
Brooks, M. A., and de Silva, J. A. F. (1975). *Talanta*, **22**, 844.
Brooks, M. A., and Hackman, M. R. (1975). *Anal. Chem.*, **47**, 2059.
Brooks, M. A., de Silva, J. A. F., and D'Arconte, L. (1973a) *J. Pharm. Sci*, **62**, 1395.
Brooks, M. A., de Silva, J. A. F., and Hackman, M. R. (1973b). *Am. Lab.*, **5**(9), 23.
Brooks, M. A., de Silva, J. A. F., and Hackman, M. R. (1973c). *Anal. Chim. Acta*, **64**, 165.
Brown, M. W., Porter, G. S., and Williams, A. R. (1974). *J. Pharm. Pharmacol*, **26** (Suppl.), 95.
Browne, J. T. (1979) In W. F. Smyth (ed.), *Polarography of Molecules of Biological Significance*, Academic Press, London, p. 111.
Buchanan, G. S., and Griffith, J. E. (1963). *J. Electroanal. Chem.*, **5**, 204.
Burmicz, J. S., Smyth, W. F., and Palmer, R. F. (1976). *Analyst*, **101**, 986.
Capitano, L. (1979). *J. Pharm. Sci.*, **68**, 311.
Caplis, M. E. (1970). *PhD Thesis*, Purdue University, West Lafayette, IN.
Carruthers, C. (1976). *Anal. Chim. Acta*, **86**, 273.
Cass, A. E. G., Davis, G., Francis, G. D., Hill, H. A. O., Aston, W. J., Higgins, I. J., Plotkin, E. V., Scott, L. D. L., and Turner, A. P. F. (1984). *Anal. Chem.*, **56**, 667.
Castels, M. A., Moore, B. L., and Ward, S. R. (1989). *Anal. Chem.*, **61**, 2534.
Caudill, W. L., Novotny, M. V., and Wightman, R. M. (1983). *J. Chromatogr.*, **261**, 415.
Chaney, E. N., and Baldwin, R. P. (1985). *Anal. Chim. Acta*, **176**, 105.
Chastel, O., Kauffmann, J.-M., Patriarch, G. J., and Christian, G. D. (1989). *Anal. Chem.*, **61**, 170.
Chastel, O., Kauffmann, J.-M., Patriarch, G. J., and Christian, G. D. (1990). *Talanta*, **37**, 213.
Chatten, L. G., and Amankwa, L. (1985). *Analyst*, **110**, 1369.
Chiang, L., James, B. D., and Magee, R. J. (1989). *Mikrochim. Acta*, **II**, 149.
Chiang, L., James, B. D., Rumble, J., and Magee, R. J. (1988). *Mikrochim. Acta*, **III**, 175.
Chou, P. P., Jaynes, P. K., and Bailey, J. L. (1985). *Clin. Chem..*, **31**, 880.
Cimbura, G., and Gupta, R. C. (1965). *J. Forensic Sci*, **10**, 228.
Clark, G. J., Goodin, R. R., and Smiley, J. W. (1985). *Anal. Chem.*, **57**, 2223.
Clifford, J. M., and Smyth, W. F. (1973). *Fresnius' Z. Anal. Chem.*, **264**, 149.
Clifford, J. M., and Smyth W. F. (1974). *Analyst*, **99**, 241.
Clifford, J. M., Smyth, M. R., and Smyth, W. F. (1974). *Fresnius' Z. Anal. Chem.*, **272**, 198.
Colas, A., Lerenard, M., Pontvianne, M., and Simons, S. (1964). *Phytopharmacie*, **13**, 3.
Concialini, V., Lippolis, M. T., and Galletti, G. C. (1989). *Analyst*, **114**, 1617.
Connor, E. F., Cheng, S. W. T., and North, W. G. (1989). *J. Chromatogr.*, **491**, 240.
Creech, G., Johnson, R. T. and Stoffer, J. O., (1982), *J. Chromatogr. Sci.* **20**, 67.
Csejka, D. A., Nakos, S. T., and Dubord, E. N. (1975). *Anal. Chem.*, **47**, 322.
Daniels, S. L., and Vanderwielen, A. J. (1981), *J. Pharm. Sci.*, **70**, 211.
Davidek, J., and Janicek, G. (1961). *Experientia*, **17**, 437.
Davidek, J., and Seifert, J. (1971). *Nahrung*, **15**, 691.
Davidson, I. E. (1979). In W. F. Smyth (ed.), *Polarography of Molecules of Biological Significance*, Academic Press, London, p. 128.
Davidson, I. E., and Smyth, W. F. (1977). *Anal. Chem.*, **49**, 1195.
Davidson, I. E., and Smyth, W. F. (1979). *Anal. Chem.*, **51**, 2127.
de Silva, J. A. F., Strojny, N., and Munnon, N. (1973). *Anal. Chim. Acta*, **66**, 23.
de Silva, J. A. F., Bekersky, I., and Brooks, M. A. (1974a). *J. Pharm. Sci*, **63**, 1943.
de Silva, J. A. F., Bekersky, I., Brooks, M. A., Weinfeld, R. E., Glover, W., and Puglisi, C. V. (1974b) *J. Pharm. Sci.*, **63**, 1440.
de Silva, J. A. F., Puglisi, C. V., Brooks, M. A., and Hackman, M. R. (1974c). *J. Chromatogr.*, **99**, 461.
Di Corcia, A., and Marchetti, M. (1991). *Anal. Chem.*, **63**, 580.
Dietrich, A. M., Milington, D. S., and Seo, Y.-H. (1988). *J. Chromatogr.*, **436**, 229.
Diez-Caballero, R. J. B., Lopez de la Torro, L., Valentin, J. F. A., and Garcia, A. A. (1989). *Talanta*, **36**, 501.

Donat, J. R., and Bruland, K. W. (1988). *Anal. Chem.*, **60**, 240.
Dong, K. L., Kryger, L., Christensen, J. K., and Thomsen, K. N. (1991). *Talanta*, **38**, 101.
Dou, L., and Krull, I. S. (1990). *Anal. Chem.*, **62**, 2599.
Drummer, O. H., Christophidis, N., Horowitz, J. D., and Lousis, W. J. (1986). *J. Chromatogr.*, **374**, 251.
Dufay, C., Lecron, J. C., Daudon, F., and Gombert, J. (1991). *J. Chromatogr.*, **563**, 224.
Dugal, R., Caille, G., and Cooper, S. F. (1973). *Union Med. Can.*, **102**, 2491.
Duthu, G. S. (1984). *J. Liq. Chromatogr.*, **7**, 1023.
Ebdon, L. (1991). Plenary Lecture presented at 'Chemical Analysis of Natural Waters in Environmental Research,' Grange-over-Sands, UK, September 4–6.
Eberle, S. J., Hoesle, C., and Krueckeberg, C. H. R. (1969). *Ber. Kernforschungzentrum Karlsruhe (KFK)*, UF44.
Eggers, H. M., Halsall, M. B., and Heineman, W. R. (1982). *Clin. Chem.*, **28**, 1848.
Eisenberg, E. J., and Cundy, K. C. (1991). *Anal. Chem.*, **63**, 845.
Eisenberg, E. J., Conzentino, P., Liversidge, G. G., and Cundy, K. C. (1990). *J. Chromatogr.*, **530**, 65.
El-Maali, N. A., and Hassan, R. M. (1990). *Bioelectrochem. Bioenergy.*, **24**, 155.
Engst, R., and Schnaak, W. (1974). *Nahrung*, **18**, 597.
Farwell, S. O., Beland, F. A., and Greer, R. D. (1975). *Anal. Chem.*, **47**, 895.
Fernandez Alvarez, J. M., and Smyth, M. R. (1989). *Analyst*, **114**, 1603.
Fidelus, J., Zietek, M., Mikolajek, A. M., and Grochowska, Z. (1972). *Mikrochim. Acta*, **84**.
Filimonova, M. M., and Gorbunova, V. E. (1973). *Zh. Anal. Khim.*, **28**, 1184.
Fleet, B., and Jee, R. D. (1969). *Talanta*, **16**, 1561.
Fleet, B., and Jee, R. D. (1970). *J. Electroanal. Chem.*, **25**, 397.
Fleet, B., Jee, R. D., and Little, C. J. (1973). *J. Electroanal. Chem.*, **43**, 349.
Florence, T. M. (1974). *J. Electroanal. Chem.*, **52**, 115.
Fogg, A. G., and Ahmed, Y. Z. (1974). *Anal. Chim. Acta*, **70**, 241.
Fogg, A. G., and Alonso, R. M. (1988). *Analyst*, **113**, 361.
Fogg, A. G., Barros, A. A., and Cabral, J. O. (1986), *Analyst*, **111**, 831.
Forsman, U. (1983). *Anal. Chim. Acta*, **146**, 71.
Fouzder, N. B., and Fleet, B. (1979). In W. F. Smyth (ed.), *Polarography of Molecules of Biological Significance*, Academic Press, London, p. 278, 284.
Frew, J. E., and Hill, H. A. O. (1987) *Anal. Chem.*, **59**, 933A.
Frew, J. E., Bayliff, S. W., Gibbs, P. N. B., and Green, M. J. (1989). *Anal. Chim. Acta*, **224**, 39.
Gao, A., Li, P, Wang, G., and Zhao, Z. (1990). *Anal. Chim. Acta*, **241**, 137.
Ge, H., Wallace, G. G., and O'Halloran, R. A. J. (1990). *Anal. Chim. Acta*, **237**, 149.
George, G. M., and Morrison, J. L. (1971). *J. Assoc. Off. Agric. Chem*, **54**, 80.
Gilbert, D. D., and Mann, J. M. (1973). *Int. J. Environ. Anal. Chem.*, **2**, 221.
Glennon, J. D., Woulfe, M. R., Senior, A. T., and Ni Choileain, N. (1989). *Anal. Chem.*, **61**, 1474.
Goicolea, M. A., Arranz, J. F., Barrio, R. J., and Gomez de Balugeraz, A. (1991). *Fresenius' J. Anal. Chem.*, **339**, 166.
Goldmsith, J. A., Jenkins, H. A., Grant, J., and Smyth, W. F. (1973). *Anal. Chim. Acta*, **66**, 427.
Gonzalez, E., and Hernandez, L. (1991). *Fresenius' J. Anal. Chem.*, **339**, 187.
Goto, M., Sakura, E., and Ishii, D. (1982). *J. Chromatogr.*, **238**, 357.
Grier, R. A., and Andrews, R. W. (1981) *Anal. Chim. Acta*, **124**, 333.
Groves, J. A. (1976). *PhD Thesis*, London University.
Guette, C., and Rips, R. (1988). In *ElectroFinnanalysis, 6–9 June, 1988, Turku, Finland, Scientific Programme Abstracts*, No. P8B, 37.
Habeeb, A. H. R., Procopio, J. R., and Hernandez, L. H. (1989). *Microchem. J.*, **39**, 247.
Hackman, M. R., Brooks, M. A., de Silva, J. A. F., and Ma, T. S. (1974). *Anal. Chem.*, **46**, 1075.

Hagen, A., Mattusch, J., and Werner, G. (1991). *Fresenius' J. Anal. Chem.*, **339**, 26.
Halbert, M. K., and Baldwin, R. P. (1985). *J. Chromatogr.*, **345**, 4B.
Hall, D. A., Berry, D. M., and Schneider, C. J. (1977). *J. Electroanal. Chem.*, **80**, 155.
Halls, D. J., Townshend, A., and Zuman, P. (1968). *Analyst*, **93**, 219.
Halvorsen, S., and Jacobsen, E. (1972). *Anal. Chim. Acta.*, **57**, 127.
Hart, J. P. (1986). In M. R. Smyth and J. G. Vos (eds), *Electrochemistry, Sensors and Analysis* (*Analytical Chemistry Symposium Series*, Vol. 25), Elsevier, Amsterdam, pp. 355, 361.
Hart, J. P., ands Jordan, P. H. (1989). *Analyst*, **114**, 1633.
Hart, J. P., Smyth, W. F., and Birch, B. J. (1976). *Proc. Anal. Div. Chem. Soc.*, **13**, 336.
Hart, J. P., Wring, S. A., and Morgan, I. C. (1989). *Analyst*, **114**, 933.
Hasebe, K., Hikima, S., Kakizaki, T., Iwashimizu, T., and Aoki, K. (1990). *Analyst*, **115**, 413.
Hashem, A., and Lubczyk, B. (1991). *J. Chromatogr.*, **563**, 216.
Heaton, R. C., and Laitinen, H. A. (1974). *Anal. Chem.*, **46**, 547.
Heineman, W. R., and Halsall, H. B. (1985). *Anal. Chem.*, **57**, 1321A.
Hernandez, L., Zapardiel, A., Lopez, J. A. P., and Rodriguez, V. (1986). In Smyth, M. R., and Vos, J. G. (eds), *Electrochemistry, Sensors and Analysis*, (*Analytical Chemistry Symposium Series*, Vol. 25), Elsevier, Amsterdam, p. 385.
Hernandez, L., Zapardiel, A., Lopez, J. A. P., and Bermejo, E. (1987). *Analyst*, **112**, 1149.
Hernandez, L., Hernandez, P., Lorenzo, E., Gonzalez, C., and Gonzalez, I. (1990). *Fresenius' Z. Anal. Chem.*, **336**, 222.
Hernandez, P., Lorenzo, E., and Hernandez, L. (1990). *Anal. Chim. Acta*, **238**, 383.
Hernandez Mendez, J., Carabias Martinez, R., Gonzalez Lopez, M. E., and Cuesta Gonazles, M. (1985). *Anal. Chim. Acta*, **176**, 121.
Hild, J., Schulte, E., and Thier, H. P. (1978). *Chromatographia*, **11**, 397.
Hon-Nami, M., and Hanya, T. (1978). *J. Chromatogr.*, **161**, 205.
Hopes, T. M. (1966). *J. Assoc. Off. Anal. Chem.*, **49**, 840.
Hou, W., and Wang, E. (1989). *Analyst*, **114**, 699.
Hou, W., and Wang, E. (1990a). *Talanta*, **37**, 841.
Hou, W., and Wang, E. (1990b). *Analyst*, **115**, 139.
Hou, W., and Wang, E. (1990c). *Anal. Chim. Acta*, **239**, 29.
Hua, C. (1988). *Talanta*, **35**, 597.
Iriyama, K., Yoshiura, M., and Iwamoto, T. (1985). *J. Liq. Chromatogr.*, **8**, 333.
Ivaska, A., and Leppinen, J. (1986). *Talanta*, **33**, 801.
Jacobs, W. A., and Kissinger, P. T. (1982). *J. Liq. Chromatogr.*, **5**, 881.
Jacobsen, E., and Jacobsen, T. V. (1971). *Anal. Chim. Acta*, **55**, 293.
Jacobsen, E., Jacobsen, T. V., and Rojahn, T. (1973). *Anal. Chim. Acta*, **64**, 473.
Jan, M. R., and Smyth, W. F. (1984a). *Analyst*, **109**, 1187.
Jan, M. R., and Smyth, W. F. (1984b). *Analyst*, **109**, 1483.
Jin, Z. L., and Rappaport, S. M. (1983). *Anal. Chem.*, **55**, 1778.
Jinkui, A., and Neeb, R. (1990a). *Fresenius' J. Anal. Chem.*, **338**, 905.
Jinkui, A., and Neeb, R. (1990b). *Fresenius' J. Anal. Chem.*, **338**, 34.
Johansson, B.-L., (1983). *Acta Univ. Ups.*
Johansson, B.-L., and Persson, B. (1978). *Anal. Chim. Acta*, **102**, 121.
Jonsson, G., Gorton, L., and Pettersson, L. (1989). *Electroanalysis*, **1**, 49.
Kadish, K. M., and Spiehler, V. R. (1975). *Anal. Chem.*, **47**, 1714.
Kalvoda, R. (1984). *Anal. Chim. Acta*, **162**, 197.
Kato, S., and Dryhurst, G. (1973). *J. Electroanal. Chem.*, **62**, 415.
Kawanishi, T., Togawa, T., Ishigami, A., Tanabe, S., and Imanari, T. (1984). *Bunseki Kagaku*, **33**, E295.
Kemula, W., and Kreminska, K. (1960). *Chem. Anal. (Warsaw)*, **5**, 611.
Kester, P. E., and Danielson, N. D. (1984). *Chromatographia*, **18**, 125.
Koen, J. G., and Huber, J. F. K. (1970). *Anal. Chim. Acta*, **51**, 303.

Kreuzig, F., and Frank, J. (1981). *J. Chromatogr.*, **218**, 615.
Krull, I. S., Ding, X.-D., Selavka, C., Bratin, K., and Forcier, G. (1984). *J. Forensic Sci.*, **29**, 449.
Kuchumova, N. A., Bepuzo, C. H., and Mamomova, R. (1963). *Tr. Vses. Nauchno-Issled. Inst. Pererab. Ispol'z Topl.*, **12**, 237.
Kuftinec, A. (1978). *Acta Pharm. Jugosl.* **28**(1), 7.
Kuwahara, S. K., Brubaker, W. F., Williams, E., Tripp, S., Douglas, F. L., and Kotakae, A. N. (1990). *J. Chromatogr.*, **534**, 260.
La Course, W. R., Krull, I. S., and Bratin, K. (1985). *Anal. Chem.*, **57**, 1810.
La Course, W. R., Jackson, W. A., and Johnson, D. C. (1989). *Anal. Chem.*, **61**, 2466.
Lau, O.-W., Luk, S.-F., and Cheung, Y.-M. (1989). *Analyst*, **114**, 1047.
Lawrence, J. F., Iverson, F., Hanekamp, H. B., Bos, P., and Frei, R. W. (1981). *J. Chromatogr.*, **212**, 245.
Le Bel, G. L., Williams, D. T., Griffiths, G., and Benoit, E. M. (1979). *J. Assoc. Off. Anal. Chem.*, **62**, 241.
Lelchien, A., Valenta, P., Nürnberg, H. W., and Patriarche, G. (1981) *Fresenius' Z. Anal. Chem.*, **306**, 156.
Lewis, J. Y., Zodda, J. P., Deutsch, E., and Heineman, W. R. (1983). *Anal. Chem.*, **55**, 708.
Liu, K. E., and Abruna, H. D. (1989). *Anal. Chem.*, **61**, 2599.
Locniskar, A., and Greenblatt, D. J. (1986). *J. Chromatogr.*, **374**, 215.
Lorenzeo, E. and Hernandez, L. (1987). *Anal. Chim. Acta*, **201**, 275.
Lu, J., Jin, W., Wang, S., and Sun, T. (1990a). *J. Electroanal. Chem.*, **291**, 49.
Lu, J., Jin, W., and Wang, S. (1990b). *Anal. Chim. Acta*, **238**, 375.
Lund, H. (1958). *Acta Chem. Scand.*, **12**, 1444.
Luo, D.-B. (1986). *Anal. Chim. Acta*, **189**, 277.
MacCrehan, W. A., Durst, R. A., and Bellama, J. M. (1977). *Anal. Lett.*, **10**, 1175.
Mairesse-Ducarmois, C. A., Patriarche, G. J., and Vandenbalck, J. L. (1975). *Anal. Chim. Acta*, **76**, 299.
Mairesse-Ducarmois, C. A., Patriarche, G. J., and Vandenbalck, J. L. (1976). *Anal. Chim. Acta*, **84**, 47.
Mairesse-Ducarmois, C. A., Patriarche, G. J., and Vandenbalck, J. L. (1977). *Anal. Chim. Acta*, **88**, 47.
Malaiyandi, M., MacDonald, S. A., and Barrette, J. P. (1969). *J. Agric. Food Chem.*, **17**, 51.
Marko-Varga, G. A., (1989). *Anal. Chem.*, **61**, 831.
Martens, D. A., and Frankenberger, W. T. (1991). *Talanta*, **38**, 245.
Maruyama, M., and Furaya, T. (1957). *Bull. Chem. Soc. Jpn.*, **30**, 659.
Maskarinec, M. P., Manning, D. L., Harvey, R. W., Griest, W. H., and Tomkins, B. A., (1984). *J. Chromatogr.*, **302**, 51.
Matsueda, T., Osaki, Y., and Shigee, S. (1982). *Bunseki Kagaku*, **31**, 59.
McCurdy, E. J., Lange, J. D., and Haygarth, P. M. (1991). Poster presentation at 'Chemical Analysis of Natural Waters in Environmental Research,' Grange-over-Sands, UK, September 4–6.
McKone, C. E., Byast, T. H., and Hance, R. J. (1972). *Analyst*, **97**, 653.
Meldrum, B. S., Smyth, M. R., Smyth, W. F., and Clifford, J. M. (1976). *Psychopharmacology*, **51**, 59.
Metrohm (1987). *Metrohm Applic. Bull.*, No. 147, January.
Meyer, S. D., and Tallman, D. E. (1983). *Anal. Chim. Acta*, **146**, 227.
Miranda Ordieres, A. J., Costa Garcia, A., Tunon Blanco, P., and Smyth, W. F. (1987). *Anal. Chim. Acta*, **202**, 141.
Mohammed, Y. H., Veening, H., and Dayton, D. A. (1981). *J. Chromatogr.*, **226**, 471.
Momberg, A., Carrera, M. E., Von Bayer, D., Bruhn, C., and Smyth, M. R. (1984). *Anal. Chim. Acta*, **159**, 119.
Moncrieff, J. (1990). *J. Chromatogr.*, **529**, 194.

Moreira, J. C., Law, G. F. J., and Fogg, A. G. (1989). *Analyst*, **114**, 1607.
Morita, K., Fukamachi, K., and Tokiwa, H. (1982). *Bunseki Kagaku*, **31**, 255.
Msimanga, H. Z., and Sturrock, P. E. (1990). *Anal. Chim. Acta*, **62**, 2134.
Munoz, E., Camacho, L., Avila, J. L., and Garcia-Blanco, F. (1989). *Analyst*, **114**, 1611.
Nagels, L. J., and Creten, W. L. (1985). *Anal. Chem.*, **57**, 2706.
Nangniot, P. (1960). *Bull. Inst. Agron. Stn. Rech. Gembloux*, **28**, 365, 373, 381.
Nangniot, P. (1964). *Anal. Chim. Acta*, **31**, 166.
Nangniot, P. (1970). In J. Duculot (ed.), *La Polarographie en Agrononmie et en Bioligie*, Gembloux, Belgium, p, 311.
Nangniot, P., Zenon-Rouland, L., and Berlemont-Frennet, M. (1978). *Analusis*, **6**, 273.
Nimmo, M. (1991). Poster presentation at 'Chemical Analysis of Natural Waters in Environmental Research,' Grange-over-Sands, UK, September 4–6.
Nimmo, M., Van den Berg, C. M. G., and Brown, J. (1989). *Mar. Chem.*, **29**, 57.
Nygren, O., Vaughan, G. T., Florence, T. M., Morrison, G. M. P., Warner, I. M., and Dale, L. S. (1990). *Anal. Chem.*, **62**, 1637.
Oates, D. M., and Jorgenson, J. W. (1989). *Anal. Chem.*, **61**, 432.
Oelschlager, H., Volke, J., Lim, G. T., and Frank, V. (1966). *Arzneim.-Forsch.*, **16**, 82.
Oelschlager, H., Volke, J., and Lim, G. T. (1967). *Arzneim.-Forsch.*, **17**, 637.
Oelschlager, H., Volke, J., and Lim, G. T. (1969). *Arch. Pharm. (Berlin)*, **302**, 241.
Okuda, Y., (1963). *Rev. Polarogr., Kyoto*, **11**, 197.
Olefirowicz, T. M., and Ewing, A. G., (1990). *J. Neurosci. Methods*, **34**, 11.
Opheim, L. N., (1977). *Anal. Chim. Acta*, **89**, 225.
Palmisano, F., and Zambonin, P. G. (1984). *Ann. Chim. (Rome)*, **74**, 633.
Peled, D., Yarnitzky, C., and Smyth, W. F. (1987). *Analyst*, **112**, 959.
Purnell, C. J., and Warwick, C. J. (1980). *Analyst*, **105**, 861.
Rao, A. (1978). *Indian J. Pharm. Sci.*, **40**, 74.
Rappaport, S. M., Jin, Z. L., and Xu, X. B. (1982). *J. Chromatogr.*, **240**, 145.
Reif, V. D., and Deangelis, N. J. (1983). *J. Pharm. Sci.*, **72**, 1330.
Ribes, A. J., and Osteryoung, J. (1990). *Anal. Chem.*, **62**, 2632.
Rocklin, R. D., and Johnson, E. L. (1983). *Anal. Chem.*, **55**, 4.
Rodriguez, J. R. B., Costa Garcia, A., and Tunon Blanco, P. (1989). *Analyst*, **114**, 939.
Rodriguez, J. R. B., Diaz, V. C., Garcia, A. C., and Tunon Blanco, P. (1990). *Analyst*, **115**, 209.
Roos, R. W. (1978). *J. Pharm. Sci.*, **67**, 1735.
Roschig, M., and Matschiner, H. (1967). *Chem. Tech (Berlin)*, **19**, 103.
Roston, D. A., Schoup, R. E., and Kissinger, P. T. (1982). *Anal. Chem.*, **54**, 1417A.
Rowe, R. R., and Smyth, M. R. (1979). In W. F. Smyth (ed.), *Polarography of Molecules of Biological Significance*, Academic Press, London, p. 249.
Sadee, W., and Van der Kleijn, E. (1971), *J. Pharm. Sci.*, **60**, 135.
Samuelsson, R., and Osteryoung, J. (1980). *Anal. Chem.*, **52**, 2215.
Samuelsson, R., and Osteryoung, J. (1981) *Anal. Chim. Acta*, **123**, 97.
Santos, L. M., and Baldwin, R. P. (1986). *Anal. Chem.*, **58**, 848.
Santos, L. M., and Baldwin, R. P. (1987a). *J. Chromatogr.*, **414**, 161.
Santos, L. M., and Baldwin, R. P. (1987b). *Anal. Chem.*, **59**, 1766.
Santos, L. M., and Baldwin, R. P. (1988). *Anal. Chim. Acta*, **206**, 85.
Sauberlich, H., Woggon, H., and Jehring, H. (1969). *Fresenius' Z. Anal. Chem.*, **246**, 185.
Schleich, C., and Henze, G. (1990). *Fresenius' Z. Anal. Chem.*, **338**, 145.
Schwartz, R. S., and David, K. O. (1985). *Anal. Chem.*, **57**, 1362.
Selavka, C. M., Krull, I. S., and Lurie, I. S. (1985), *J. Chromatogr. Sci.*, **23**, 499.
Selavka, C. M., Krull, I. S., and Bratin, K. (1986). *J. Pharm. Biomed. Anal.* **4**, 83.
Senguen, F. I., and Oelschlager, H. (1975). *Arch. Pharm. (Berlin)*, **308**, 720.
Senkowski, B. Z., Levin, M. S., Urbigkit, J. R., and Wollish, E. G. (1964). *Anal. Chem.*, **36**, 1991.

Sestakova, I., and Kopanica, M. (1988). *Talanta*, **35**, 816.
Shibanoki, S., Weinberger, S. B., Ishikawa, K., and Martinez, J. L. (1990). *J. Chromatogr.*, **532**, 249.
Shinozuka, N., Suzuki, H., and Hayano, S. (1972). *Bunseki Kagaku*, **21**, 517.
Sithole, B. P., and Guy, R. D. (1986). *Talanta*, **33**, 95.
Smyth, M. R., and Frischkorn, C. G. B. (1980). *Fresenius' Z. Anal. Chem.*, **301**, 220.
Smyth, M. R., and Osteryoung, J. G. (1977). *Anal. Chem.*, **49**, 2310.
Smyth, M. R., and Osteryoung, J. G. (1978). *Anal. Chim. Acta*, **96**, 335.
Smyth, M. R., Smyth, W. F., Palmer, R. F., and Clifford, J. M. (1976a). *Anal. Chim. Acta*, **86**, 185.
Smyth, M. R., Smyth, W. F., Palmer, R. F., and Clifford, J. M. (1976b). *Analyst*, **101**, 469.
Smyth, M. R., Beng, T. S., and Smyth, W. F. (1977a). *Anal. Chim. Acta*, **92**, 129.
Smyth, M. R., Osteryoung, J. G., Rowley, P. G. (1977b). In *Proc. 174th ACS Meeting, Chicago*, American Chemical Society, Washington, DC. **89**, 1978.
Smyth, M. R., Smyth, W. F., and Clifford, J. M. (1977c). *Anal. Chim. Acta*, **94**, 119.
Smyth, W. F. (1984). Unpublished work.
Smyth, W. F. (1987). *CRC Crit. Rev. Anal. Chem.* **18**, 155.
Smyth, W. F. (1989). *The Application of Modern Electrochemical and Chromatographic Techniques of Analysis to the Determination of Pharmaceuticals of Significance in Zambia*, Research Report, University of Zambia.
Smyth, W. F., and Hassanzadeh, M. (1976). *Fresenius' Z. Anal. Chem.*, **280**, 299.
Smyth, W. F., and Ivaska, A. (1985). *Analyst*, **110**, 1377.
Smyth, W. F., and Yarnitzky, C. (1988). *Int. J. Pharm.*, **42**, 235.
Smyth, W. F., Svehla, G., and Zuman, P. (1970a). *Anal. Chim. Acta*, **51**, 463.
Smyth, W. F., Svehla, G., and Zuman, P. (1970b). *Anal. Chim. Acta*, **52**, 129.
Smyth, W. F., Svehla, G., and Zuman, P. (1971). *J. Electroanal. Chem.*, **30**, 101.
Smyth, W. F., Jenkins, T. C., Siekiera, J. and Baydar, A. (1975a). *Anal. Chim. Acta*, **80**, 233.
Smyth, W. F., Watkiss, P., Burmicz, J. S., and Hanley, H. O. (1975b). *Anal. Chim. Acta*, **78**, 81.
Stara, V., and Kopanica, M. (1988). *Anal. Chim. Acta*, **208**, 231.
Stewart, J. T., and Carter, H. K. (1986). *J. Chromatogr.*, **380**, 177.
Stitz, A., and Buchberger, W. (1991). *Fresenius' J. Anal. Chem.*, **339**, 55.
Supin, G. S., and Budnikov, G. K. (1973). *Zh. Anal. Khim.*, **28**, 1459.
Svagrova, I., Stulik, K., Pacakova, V., Caliceti, P., and Veronese, F. M. (1991). *J. Chromatogr.*, **563**, 95.
Swagzdis, J. E., Gifford, R., and Mico, B. A. (1985). *J. Chromatogr.*, **345**, 203.
Takano, B., McKibben, M. A., and Barnes, H. L. (1984). *Anal. Chem.*, **56**, 1594.
Tanaka, S., Sugawara, K., and Taaga, M. (1990). *Talanta*, **37**, 1001.
Telting-Diaz, M., Miranda Ordieres, A. J., Costa Garcia, A., Tunon Blanco, P., Diamond, D., and Smyth, M. R. (1990). *Analyst*, **115**, 1215.
Thoma, K., and Klimek, R. (1980). *Dtsch. Apoth.-Ztg.*, **142**, 1967.
Thomas, M. B. (1986). *Diss. Abstr. Int. B.*, **46**(12), Pt. 1, 4216.
Thompson, J. M. (1965). *MSc Thesis*, University of Birmingham, Birmingham, AL.
Thomsen, K. N., Kryger, L., and Baldwin, R. P. (1988). *Anal. Chem.*, **60**, 151.
Thrivikraman, K. U., Refshange, C., and Adams, R. N. (1974). *Life Sci.*, **15**, 1335.
Tocksteinova, Z., and Kopanica, M. (1987). *Anal. Chim. Acta*, **199**, 77.
Tolbert, A. M., and Baldwin, R. P. (1989). *Electroanalysis*, **1**, 389.
Torrance, K., and Gatford, C. (1987). *Talanta*, **34**, 939.
Urbansky, J., Czerwinski, W., Janicka, K., Majewska, F., and Zowall, H. (1977), *Handbook of Analysis of Synthetic Polymers and Plastics*, Ellis Horwood, Chichester.
Vallon, J. J., Badinard, A., and Bichov, C. (1975). *Anal. Chim. Acta*, **78**, 93.
Van Beeumen, J., and Deley, J. (1971). *Anal. Biochem.*, **44**, 254.

van Bennekom, W. P., Reenwijk, H. J. E. M., and Schute, J. B. (1975). *Anal. Chim. Acta*, **74**, 387.
Van den Berg, G. M. G., and Huang, Z. Q. (1984). *J. Electroanal. Chem.*, **177**, 269.
Van Kerchove, C., Bontempts, R., and Schoenmakers, A. (1982). *J. Pharm. Pharmacol.*, **34**, 420.
Vigh, G., Varga-Puchony, A., Hlavay, J., Petro-Turza, M., and Szarfoldi-Szalma, I. (1980). *J. Chromatogr.*, **193**, 432.
Vire, J.-C., Maali, N. A. E., Patriarche, G. J., and Christian, G. D. (1988). *Talanta*, **35**, 997.
Visconti, A., Bottalico, A., Palmisano, F., and Zambonin, P. G. (1984). *Anal. Chim. Acta*, **159**, 111.
Vree, T. B. (1978). *J. Chromatogr.*, **146**, 103.
Wall, G. M., and Baker, J. K. (1989). *J. Chromatogr.*, **491**, 151.
Walters, C. L., Johnson, E. M., and Ray, N. (1970). *Analyst*, **95**, 485.
Walton, A. P., Wei, G.-T., Liang, Z., Michel, R. G., and Morris, J. B. (1991). *Anal. Chem.*, **63**, 232.
Wang, J., and Freiha, B. A. (1983). *Anal. Chem.*, **55**, 1285.
Wang, J., and Mahmoud, J. S. (1986). *Anal. Chim. Acta*, **186**, 31.
Wang, J., and Varughese, K. (1987). *Anal. Chim. Acta*, **199**, 185.
Wang, J., Farias, P. A. M., and Mahmoud, J. S. (1985a). *Anal. Chim. Acta*, **171**, 195.
Wang, J., Farias, P. A. M., and Mahmoud, J. S. (1985b). *Anal. Chim. Acta*, **172**, 57.
Wang, J., Mahmoud, J. S., and Farias, P. A. M. (1985c). *Analyst*, **110**, 855.
Wang, J., Luo, D. B., Farias, P. A. M., and Mahmoud, J. S. (1985d). *Anal. Chem*, **57**, 158.
Wang, J., Peng, T., and Lin, S. M. (1986). *Bioelectrochem. Bioenerg.*, **15**, 147.
Wang, J., Tuzhi, P., and Varughese, K. (1987a). *Talanta*, **34**, 561.
Wang, J., Zadeii, J., and Lin, M. S. (1987b). *J. Electroanal. Chem.*, **237**, 281.
Wang, J., Li, R., and Lin, M. S. (1989). *Electroanalysis*, **1**, 151.
Wang, J., Sun, C., and Jin, W. (1990). *J. Electroanal. Chem.*, **291**, 59.
Warner, A. M., and Weber, S. G. (1989). *Anal. Chem.*, **61**, 2664.
Warwick, C. J., Bagon, D. A., and Purnell, C. J. (1981). *Analyst*, **106**, 676.
Watson, A., and Svehla, G. (1974). *Proc. Soc. Anal. Chem.*, **11**, 163.
Watson, A., and Sevehla, G. (1975a). *Analyst*, **100**, 489.
Watson, A., and Sevehla, G. (1975b). *Analyst*, **100**, 573.
Watson, A., and Sevehla, G. (1975c). *Analyst*, **100**, 584.
Webb, L. E., and Johnson, R. C. (1986). *Clin. Biochem.*, **19**, 212.
Wehmeyer, K. R., Halsall, H. B., and Heineman, W. R. (1982). *Clin. Chem.*, **28**, 1968.
Wehmeyer, K. R., Halsall, H. B., Heineman, W. R., Volle, C. P., and Cheu, I. (1986). *Anal. Chem.*, **58**, 135.
Welch, L. E., La Course, W. R., Mead, D. A., and Johnson, D. C. (1990). *Talanta*, **37**, 377.
Wieck, H. J., Hieder, G. H., and Yacynych, A. M. (1984). *Anal. Chim. Acta*, **158**, 137.
Williams, A. F., and Murray, W. J. (1966). *Nature (London)* **210**, 816.
Wring, S. A., Hart, J. P., and Birch, B. J. (1989). *Analyst*, **114**, 1571.
Wu, W. S., Stoyanoff, R. E., Szklar, R. S., Gaind, V. S., and Rakanovic, M. (1990). *Analyst*, **115**, 801.
Yarnitzky, Ch., and Smyth, W. F. (1991). *Int. J. Pharma.*, **75**, 161.
Zadeii, J. M., Marioli, J., and Kuwana, T. (1991). *Anal. Chem.*, **63**, 649.
Zapardiel, A., Perez Lopez, J. A., Bermejo, E., and Hernandez, L. (1988). *Fresenius' Z. Anal. Chem.*, **330**, 707.
Zapardiel, A., Perez Lopez, J. A., Bermejo, E., Hernandez, L., and Valenciano, M. J. (1990). *Microchem. J.*, **41**, 10.
Zapardiel, A., Perez Lopez, J. A., Bermejo, E., Hernandez, L., and Chicharro, M. (1991). *Anal. Chim. Acta*, **244**, 49.
Zhang, H., Vire, J.-C., Patriarche, G. J., and Wollast, R. (1988). *Anal. Lett.*, **21**, 1409.
Zhang, H., Wollast, R., Vire, J.-C., and Patriarche, G. J. (1989). *Analyst*, **114**, 1597.

Zhao, J., and Jin, W. (1988). *J. Electroanal. Chem.*, **256**, 181.
Zhao, J. and Jin, W. (1989). *J. Electroanal. Chem.*, **267**, 271.
Zhu, S.-M., and Brajter-Toth, A. (1990). *Anal. Chim. Acta*, **237**, 305.
Zoulis, N. E., Nikolelis, D. P., and Efstathiou, C. E. (1990). *Analyst*, **115**, 291.
Zuman, P., and Perrin, C. L. (1969). *Organic Polarography*, 4th edn, Wiley, London, New York, p. 161.

Index

27 (1973 – 1991)